Toward High School Biology

Understanding Growth in Living Things

Student Edition

Claire Reinburg, Director
Rachel Ledbetter, Managing Editor
Deborah Siegel, Associate Editor
Amanda Van Beuren, Associate Editor
Donna Yudkin, Book Acquisitions Manager

Art and Design
Will Thomas Jr., Director

Printing and Production
Catherine Lorrain, Director

National Science Teachers Association
David L. Evans, Executive Director
David Beacom, Publisher

1840 Wilson Blvd., Arlington, VA 22201
www.nsta.org/store
For customer service inquiries, please call 800-277-5300.

20 19 18 17 4 3 2 1

Library of Congress Cataloging-in-Publication Data
Names: National Science Teachers Association.
Title: Toward high school biology : understanding growth in living things.
Other titles: Understanding growth in living things
Description: Student edition. | Arlington, VA : National Science Teachers Association, 2017.
Identifiers: LCCN 2017021525 (print) | LCCN 2017025660 (ebook) | ISBN 9781681404448 (e-book) | ISBN 9781681404431 (print)
Subjects: LCSH: Growth--Juvenile literature.
Classification: LCC QH511 (ebook) | LCC QH511 .T69 2017b (print) | DDC 571.8--dc23
LC record available at https://lccn.loc.gov/2017021525

Contents

Chapter 1 — 1

Lesson 1.1—Changes in Living and Nonliving Things 2
Lesson 1.2—Detecting New Substances 6
Lesson 1.3—Making New Substances From Other Substances 24
Lesson 1.4—Using Models to Think About Atoms and Molecules 34
Lesson 1.5—Using Models to Represent Chemical Reactions 45
Lesson 1.6—Representing Chemical Reactions That Produce Large Molecules 61

Chapter 2 — 69

Lesson 2.1—Chemical Reactions and Mass 70
Lesson 2.2—Sealed Containers and Total Mass. 77
Lesson 2.3—Opened Containers and Measured Mass 86

Chapter 3 — 101

Lesson 3.1—The "Stuff" That Makes Up Plants 102
Lesson 3.2—Carbohydrates That Make Up Plants 108
Lesson 3.3—Making Glucose in Plants. 116
Lesson 3.4—Making Cellulose Polymers in Plants 126
Lesson 3.5—Explaining Where the Mass of Growing Plants Comes From. 133

Chapter 4 — 141

Lesson 4.1—The "Stuff" That Makes Up Animals. 142
Lesson 4.2—Proteins in Animal Bodies and Food 148
Lesson 4.3—Explaining Animal Growth With Atoms and Molecules. 157
Lesson 4.4—Examining Explanations of Animal Growth and Repair 170
Lesson 4.5—Explaining Growth in All Living Things .176

Note: Color images of models are available on the book's Extras page at *www.nsta.org/towardhsbio.*

AAAS Project 2061 Team
Jo Ellen Roseman, Principal Investigator
Cari Herrmann-Abell, Senior Research Associate
Jean Flanagan, Research Associate
Ana Cordova, Research Assistant
Caitlin Klein, Research Assistant
Bernard Koch, Research Assistant
Mary Koppal, Communications Director
Abigail Burrows, Senior Project Coordinator

External Review Team
Marlene Hilkowitz, Temple University, Philadelphia, PA
Michele Lee, Temple University, Philadelphia, PA
Barb Neureither, Michigan State University, East Lansing, MI
Edward Smith, Michigan State University, East Lansing, MI

Consultant
Kathy Vandiver, Massachusetts Institute of Technology, Boston, MA

Acknowledgments
The *Toward High School Biology* unit has benefited immensely from the many contributions of the staff at BSCS, who worked in partnership with AAAS during the first three years of the project (2010 through 2013). In particular, we are grateful to Janet Carlson, Rhiannon Baxter, Brooke Bourdélat-Parks, Elaine Howes, Rebecca Kruse (currently at the National Science Foundation), Stacey Luce, Chris Moraine, Kathleen Roth, and Kerry Skaradzinski for their expertise and tireless efforts. Recent contributions (2014 to 2016) of Rebecca Kruse were funded by the National Science Foundation Independent Research/Development Program. Any opinions, findings, conclusions, or recommendations expressed in this publication are those of the authors and do not necessarily reflect the views of the National Science Foundation.

We would also like to thank the many excellent teachers and their students in Colorado, the District of Columbia, Maryland, and Massachusetts who participated in pilot- and field-testing the unit. Their insights and feedback have been invaluable.

Development of *Toward High School Biology* was funded by the U.S. Department of Education Institute of Education Sciences, Grant #IES-R305A100714. Any opinions, findings, conclusions, or recommendations expressed in this publication are those of the authors and do not necessarily reflect the views of the funding agency.

Toward High School Biology

Chapter 1

Student Edition

Lesson 1.1—Changes in Living and Nonliving Things

What do we know and what are we trying to find out?

Everything in the world—both things that are alive and things that aren't—is made of "stuff" that is changing all of the time. Scientists call this "stuff" matter. For example, a bicycle is matter, and when it is left outside, it becomes covered with rust, which is also matter. Matter changes when a baker makes a cake from flour, sugar, eggs, baking powder, and butter. Students mix vinegar with baking soda and change these ingredients into a foaming "volcano." Wood burns in a campfire and turns to ashes. Scientists in the lab combine different substances to create new products.

Living things are also made of matter, and most of them—including you—change as they grow and repair their bodies (see the color version of Figure 1.1 that your teacher will project). A small puppy grows into a big dog. A tiny seed sprouts a seedling that then grows into a giant sequoia tree. The skin on your finger heals around a cut. A lizard regrows a tail that it lost to a predator.

Figure 1.1. Changes in Matter

You can probably think of many more examples of how matter changes. But how do these changes happen? Do living and nonliving things change in the same way?

To answer these questions, we need to understand more about the "stuff" that makes up living and nonliving things and how it changes. In this lesson, you will observe some of these changes and begin thinking about the Key Question (there is no need to respond in writing now).

> **Key Question: How are changes in the matter that makes up living and nonliving things similar?**

Activity 1: Observing Changes in Living and Nonliving Things

In this activity, you will watch a series of videos that show different changes that occur in living and nonliving things.

Procedures and Questions

1. Your teacher will show two videos. Watch each video and record what you see happening in Table 1.1. Be sure to include any observations that suggest change is occurring.

Table 1.1. Changes in Living Things

	Observations
Video 1– Puppy Growth	
Video 2– Corn Growth	

2. Discuss these questions with your class. Use your observations from Table 1.1 to support your ideas.
   a. What matter is there before each change takes place?
   b. Do you think any new matter is being made during each change?
   c. If so, what is the new matter and where do you think it comes from?

3. Your teacher will show two more videos. Watch each video and record what you see happening in Table 1.2. Be sure to include any observations that suggest change is occurring.

Table 1.2. Changes in Nonliving and Living Things

	Observations
Video 1– Nylon Formation	
Video 2– Spider Spinning Silk	

4. Discuss these questions with your class. Use your observations from Table 1.2 to support your ideas.
   a. What matter is there before each change takes place?
   b. Do you think any new matter is being made during each change?
   c. If so, what is the new matter and where do you think it comes from?
   d. What do you think has happened to the matter that was there in the beginning?

Pulling It Together

Work on your own to answer the question. Be prepared for a class discussion.

1. After observing and discussing the changes shown in the videos, how would you answer the Key Question, **How are changes in the matter that makes up living and nonliving things similar?**

Lesson 1.2—Detecting New Substances

What do we know and what are we trying to find out?

All of the changes that you observed in Lesson 1.1—in both the living and nonliving examples—involved changes in matter. Understanding what matter is made up of is key to understanding how matter changes.

Some matter, like solid table sugar, liquid water, and oxygen gas, consists of one substance. More commonly, matter is made up of several different substances—air, for example, is a mixture mainly of nitrogen gas, oxygen gas, water vapor, and carbon dioxide gas. Likewise, the matter that composes the bodies of living organisms—dogs, trees, humans, and lizards, for example—is made up of many different substances that are organized into complex structures.

Many substances look very similar. Take water and vinegar, for example (Figure 1.2).

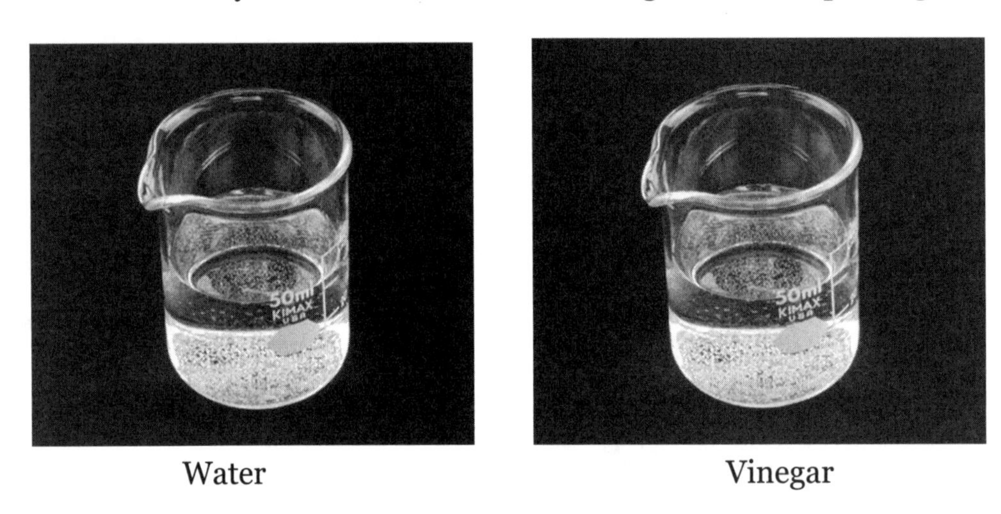

Figure 1.2. Comparing Water and Vinegar

Both are clear colorless liquids. So if you had a cup of each, how would you know which is which? If you have had experiences with vinegar, you may have noticed that it has a very strong odor, and you probably know that water doesn't have much of an odor at all. The odor of a substance is one way to tell if two substances are the same or different.

Properties like odor are called "characteristic" properties because they help define that substance. Other characteristic properties include the melting point and boiling point of a substance, whether or not a substance can be dissolved in water, and the electrical conductivity of a substance. No matter how small the sample of a substance is, what shape it is, when it is observed, or where it is located, a substance's characteristic properties will always be the same. And even though different substances may have **some** characteristic properties that are the same, no two substances have characteristic properties that are **all** the same. For this reason, scientists can use the set of characteristic properties of a substance to identify it when they don't know what it is.

In this lesson, you will explore the characteristic properties of the substances that were involved in some of the changes you saw in the videos in Lesson 1.1. This will help you make sense of what is happening during each of these changes.

Answer the Key Question to the best of your knowledge. Be prepared to share your ideas with the class.

Key Question: How do we know if a different substance has been made?

Activity 1: Investigating a Change: Hexamethylenediamine and Adipic Acid

Materials
For each team of students
Card Pack 1 (green outline)
Card Pack 2 (yellow outline)
Change Chart

In this activity, you will investigate the first of three different examples of changes in matter. You will refer to these three changes often during the unit, so it is important that you make very careful observations.

To help in your investigations, you will look at charts that represent the three changes and cards that describe the starting and ending substances that are involved. Each card provides data on the characteristic properties for a substance. You will use these cards and charts to think about how we can tell if a different substance has been made.

Procedures and Questions

1. As a whole class, brainstorm a list of words to describe what you are able to observe about substances—their physical state, appearance, or smell, for example. Your teacher will record this list on the board or chart paper so that you can use it whenever you write descriptions of substances. If you think of more ways to describe substances during the activity, add them to the list. You can use the space below to brainstorm.

2. Notice the labels at the top of each column in Table 1.3 (p. 15). In this activity, you will be writing information in the appropriate boxes in some or all of the columns. To get started, write the names for the starting substances involved in the change you are investigating. Hint: Look at the title of the table.

3. Observe the video of the starting substances and fill in whatever information you now have about the substances in the appropriate columns in rows 1 and 2 of Table 1.3.

4. Using *Card Pack 1* and the *Change Chart,* do the following:

   a. Find the cards for Hexamethylenediamine and Adipic Acid. Place them on the Starting Substances (left) side of the *Hexamethylenediamine and Adipic Acid Change Chart.*

   b. Compare the observations you recorded for starting substance 1 and starting substance 2 in Table 1.3 with the properties on the Hexamethlyenediamine and Adipic Acid cards. Fill in the table with information from the data cards. Does the information on the cards match your observations? What else can you learn about these substances from the cards?

5. Watch as your teacher presents a video of the change. Make careful observations and write detailed descriptions of what you observed during the change in row 3 of Table 1.3. Use words from the list you brainstormed in Step 1.

6. The starting substances in this investigation are too dangerous to have in the classroom. However, the ending substances are safe. Ask your teacher for pure samples of the ending substances.

7. Observe each ending substance. Record observations about the ending substance's state of matter (at room temperature), color, and smell in Table 1.3.

   When smelling a substance, remove the cap and wave your hand over the opened container to waft the odor toward your nose. Never smell a substance directly.

8. Test each ending substance's conductivity by touching both leads (metal prongs) of the conductivity meter to the substance. Record whether each substance is conductive in rows 4 and 5 of Table 1.3 in the Conductive column.

   If the bulb or diode lights or flashes, electricity is being conducted through the sample and the substance is considered conductive.

9. Place a small amount of the ending substance into a vial. Test its solubility by filling the vial halfway with water. Record whether each substance is soluble in rows 4 and 5 of Table 1.3 in the Soluble column.

10. If you observed that the substance is soluble, test its conductivity again. Record your observation for each substance in rows 4 and 5 of Table 1.3 in the Conductive When Dissolved column. If the substance is not soluble, write N/A.

11. Using *Card Pack 2*, do the following:

    a. Compare your observations of the ending substances that you recorded in Table 1.3 with the properties of the different substances shown in *Card Pack 2* for the Hexamethylenediamine and Adipic Acid investigation.

    b. Now look at the *Hexamethylenediamine and Adipic Acid Change Chart*. Match any substance card(s) that you can to this change. Place them on the Ending Substances (right) side of the chart. Write the name(s) of the ending substance(s) in the first column of rows 4 and 5 of Table 1.3.

Activity 2: Investigating a Change: Steel Wool (Iron) and Air (Oxygen)

Materials
For each team of students
Dry steel wool
Steel wool soaking in vinegar
Paper towels
Flask
Finger of a nitrile glove
Card Pack 1 (green outline)
Card Pack 2 (yellow outline)
Change Chart

In this activity, you will investigate another change using your own observations and information on the substance cards and *Change Chart.*

Procedures and Questions

1. Write the names for the starting substances involved in this change in Table 1.4 (p. 16).

2. Observe the dry steel wool. Record your observations about the steel wool's state of matter, color, and smell in the appropriate columns of Table 1.4.

3. Test the steel wool's conductivity by touching both leads of the conductivity meter to the substance. Record whether the substance is conductive in Table 1.4.

4. Place a small amount of steel wool into a vial. Test its solubility by filling the vial halfway with water. Record whether the substance is soluble in Table 1.4.

5. If you observed that the substance is soluble, test its conductivity again. Record your observation in Table 1.4. If the substance is not soluble, write N/A.

6. Steel wool has a protective coating on it that must be removed to expose the iron contained in the thin strands of steel wool. Soaking the steel wool in vinegar removes the protective coating. Vinegar is not considered a starting substance in this case, but it helps the change occur. Remove a small piece of steel wool from the vinegar. Squeeze out most of the vinegar and blot off the rest with paper towels.

7. Pull apart the strands so that more steel wool is exposed to the air and then place the steel wool in a clean, dry flask. Immediately cover the mouth of the flask with one finger cut from a nitrile glove. Take note of how the finger of the glove looks now.

8. Observe the air around you. Record your observations about the air's state of matter, color, and smell in the appropriate columns in Table 1.4. Because air is made up of several different gases, how can we find out if there is oxygen in the air?

9. Watch as your teacher shows a demonstration of a test for the presence of oxygen.

   When oxygen gas is present, a burning splint will continue to burn.

10. Using *Card Pack 1* and the *Change Chart,* do the following:

    a. Find the cards for Steel Wool (Iron) and Air (Oxygen). Place them on the Starting Substances (left) side of the *Steel Wool (Iron) and Air (Oxygen) Change Chart.*

    b. Compare the observations that you recorded for steel wool and air in Table 1.4 with the properties on the Iron and Oxygen cards. Fill in the table with information from the data cards. Does the information on the cards match your observations? What else can you learn about these substances from the cards?

11. Watch as your teacher presents a video of the change you are investigating. Make careful observations and write detailed descriptions of what you observe during the change in Table 1.4. Use words from the list you brainstormed in Step 1 of Activity 1.

12. Ask your teacher for a pure sample of the ending substance. Also note that the ending substance is being formed in your flask.

13. Observe the ending substance. Record observations about the ending substance's state of matter (at room temperature), color, and smell in Table 1.4.

14. Test the ending substance's conductivity by touching both leads (metal prongs) of the conductivity meter to the substance. Record whether the substance is conductive in Table 1.4 in the Conductive column.

15. Place a small amount of the ending substance into a vial. Test its solubility by filling the vial halfway with water. Record whether the substance is soluble in Table 1.4 in the Soluble column.

16. If you observed the ending substance to be soluble, test its conductivity again. Record whether the substance is conductive when dissolved in Table 1.4 in the Conductive When Dissolved column. If the substance is not soluble, write N/A.

17. Using *Card Pack 2,* do the following:

    a. Compare your observations of the ending substance to the properties of the different substances in *Card Pack 2* for the Steel Wool (Iron) and Air (Oxygen) investigation.

    b. Match any substance card(s) that you can to this change. Place them on the Ending Substances (right) side of the *Steel Wool (Iron) and Air (Oxygen) Change Chart.* Write down the name of the ending substance in Table 1.4.

Activity 3: Investigating a Change: Baking Soda and Vinegar

Materials
For each team of students
Vinegar
Baking soda
Teaspoon
Measuring cup
Quart-size zipper plastic bag
Card Pack 1 (green outline)
Card Pack 2 (yellow outline)
Change Charts

In this third change, it is important to keep track of all of the substances that are involved.

Procedures and Questions

1. Write the names for the starting substances involved in this change in Table 1.5 (p. 17).
2. Observe the baking soda. Record your observations about baking soda's state of matter, color, and smell in the appropriate columns in Table 1.5.
3. Test baking soda's conductivity by touching both leads of the conductivity meter to the substance. Record whether the substance is conductive in Table 1.5.
4. Place a small amount of baking soda into a vial. Test its solubility by filling the vial halfway with water. Record whether the substance is soluble in Table 1.5.
5. If you observed that the substance is soluble, test its conductivity again. Record your observation in Table 1.5 in the Conductive When Dissolved column. If the substance is not soluble, write N/A.
6. Repeat Steps 2–5 for vinegar.
7. Using *Card Pack 1* and the *Change Chart,* do the following:
   a. Find the cards for Baking Soda and Vinegar. Place them on the Starting Substances (left) side of the *Baking Soda and Vinegar Change Chart.*
   b. Compare the observations that you recorded in Table 1.5 to the properties on the Baking Soda and Vinegar cards. Fill in the table with information from the data cards. Does the information on the cards match your observations? What else can you learn about these substances from the cards?

8. Put a teaspoon of baking soda directly into the quart-size bag. Measure 25 ml of vinegar in a measuring cup.

9. Place the cup containing vinegar upright in the plastic bag. Take care that you do not spill the contents of the cup. Push as much air out of the plastic bag as possible and seal the bag. Observe how the bag looks and feels now, and record your observations in Table 1.5. Then, without opening the bag, tip over the cup of vinegar in the bag so that the vinegar mixes with the baking soda. Make careful observations of what happens in the bag during the change.

10. Write a detailed description of what happened inside the bag during the change in row 3 of Table 1.5. Write about any changes in how the bag looks and feels now compared with how it looked and felt when you completed Step 9.

11. Watch as your teacher shows a demonstration of the limewater test.

    If a colorless limewater solution turns milky white (test result is positive), it means that the limewater solution has interacted with carbon dioxide gas. In other words, the sample has tested positive for carbon dioxide gas.

12. Record your observations about ending substance 1 in Table 1.5.

13. Ask your teacher for pure samples of the ending substances 2 and 3.

14. Observe ending substances 2 and 3. For each ending substance, record observations about its state of matter (at room temperature), color, and smell in Table 1.5.

15. Test the conductivity of ending substances 2 and 3 by touching both leads (metal prongs) of the conductivity meter to the substance. Record whether each substance is conductive in Table 1.5 in the Conductive column.

16. Place a small amount of each ending substance into two separate vials. Test the solubility of each by filling the vial halfway with water. Record whether each substance is soluble in Table 1.5 in the Soluble column.

17. If you observed either of the ending substances to be soluble, test its conductivity again. Record whether the substance is conductive when dissolved in Table 1.5 in the Conductive When Dissolved column. If the substance is not soluble, write N/A.

18. Using *Card Pack 2* and the *Change Chart,* do the following:

    a. Compare your observations of the ending substances with the properties of the different substances in *Card Pack 2* for the Baking Soda and Vinegar investigation.

    b. Match any substance card(s) that you can to this change. Place them on the Ending Substances (right) side of the *Baking Soda and Vinegar Change Chart*. Write down the name of the ending substance(s) in Table 1.5.

Table 1.3. Hexamethylenediamine and Adipic Acid Investigation

Starting Substances						
Name	State of Matter (at room temperature)	Color	Smell	Conductive	Soluble	Conductive When Dissolved
1. Starting Substance 1						
2. Starting Substance 2						
What did you observe during the change?						
3.						
Ending Substances						
Name	State of Matter (at room temperature)	Color	Smell	Conductive	Soluble	Conductive When Dissolved
4. Ending Substance 1						
5. Ending Substance 2						

Table 1.4. Steel Wool (Iron) and Air (Oxygen) Investigation

Starting Substances						
Name	State of Matter (at room temperature)	Color	Smell	Conductive	Soluble	Conductive When Dissolved
1. Starting Substance 1						
2. Starting Substance 2						
What did you observe during the change?						
3.						
Ending Substance						
Name	State of Matter (at room temperature)	Color	Smell	Conductive	Soluble	Conductive When Dissolved
4. Ending Substance 1						

Table 1.5. Baking Soda and Vinegar Investigation

Starting Substances						
Name	State of Matter (at room temperature)	Color	Smell	Conductive	Soluble	Conductive When Dissolved
1. Starting Substance 1						
2. Starting Substance 2						
What did you observe during the change?						
3.						
Ending Substances						
Name	State of Matter (at room temperature)	Color	Smell	Conductive	Soluble	Conductive When Dissolved
4. Ending Substance 1						
5. Ending Substance 2						
6. Ending Substance 3						

19. With your class, fill out the table that your teacher is going to project. Then answer the following questions:

    a. Compare the properties of the substances. Do ANY of the substances have exactly the same set of properties? Give an example from data on the substance cards or from your observations to support your answer.

    b. Did substances that had a similar appearance have the same set of properties? Give an example from data on the substance cards or from your observations to support your answer.

Science Ideas

Science ideas are accepted principles or generalizations about how the world works based on a wide range of observations and data collected and confirmed by scientists. Because these science ideas are consistent with the available evidence, you are justified in applying science ideas to other relevant observations and data.

Activities 1, 2, and 3 were intended to help you understand an important idea about substances and their properties. Read the idea below. Look back through Lesson 1.2. In the space provided after the science idea, give evidence that supports the idea.

Science Idea #1: Every substance has a unique set of characteristic properties, such as color, odor, density, melting point, conductivity, solubility, and how it behaves (such as in the limewater and burning splint tests). The properties of substances can be observed or measured and used to decide if two substances are the same or different.

Evidence (from your observations or data on the substance cards):

Activity 4: Constructing Scientific Explanations—Part I

Scientists try to explain how the world works using evidence from their observations, science ideas, and logical reasoning that links the evidence and the science ideas. Throughout this unit, you too will be asked to make observations and explain them using evidence, science ideas, and logical reasoning. This activity introduces you to what a scientific explanation needs to include and how to evaluate its quality.

Procedures and Questions

1. After completing Activities 1, 2, and 3, students were asked whether a new substance was formed when hexamethylenediamine and adipic acid were mixed and to explain their claims. Eva looked at the cards for hexamethylenediamene, adipic acid, and nylon and decided that a new substance was formed.

   To help write a valid explanation of her claim, Eva organized her thinking as shown in Table 1.6. Examine the different parts of her explanation and read the description of each part.

Table 1.6. Parts of an Explanation

	What It Is	Eva's Explanation
Question	The question to be answered	
Claim	A statement or conclusion that is intended to respond to the question	*Yes, a new substance formed when hexamethylenediamine and adipic acid were mixed.*
Science Ideas	Widely accepted scientific ideas, concepts, or principles that can be used to show why the evidence supports the claim	*Every substance has its own set of properties (Science Idea #1).*
Evidence	Data that are relevant to and support the claim and can be confirmed by others Data are based on observations about the world that are either made with our senses or measured with instruments.	*Hexamethylenediamine and adipic acid are clear, colorless liquids and soluble in water.* *Nylon is a white, fibrous solid and not soluble in water.*

Eva's explanation: *Yes, a new substance formed when hexamethylenediamine and adipic acid were mixed. Every substance has its own set of properties. If two substances have different properties, they are different substances. Properties can be used to distinguish one substance from another (Science Idea #1). The most likely ending substance from the card pack, nylon, has a different set of properties than the two starting substances, hexamethylenediamine and adipic acid. For example, both hexamethylenediamine and adipic acid are clear, colorless liquids that are both soluble in water, whereas nylon is a white, fibrous solid that is not soluble in water. Because nylon has a different set of properties than the two starting substances, it is a different substance.*

Here are some questions that can help us decide whether Eva's explanation is valid. Examine Eva's explanation and answer the questions about each part. Discuss whether or not Eva's explanation answers each question, and record any ideas that come up during the discussion.

Explanation Quality Questions

- Does the **claim** respond to the question?
- Are the **science ideas** that are listed relevant to the claim?
- Does the **evidence** support the claim?
- Are the **science ideas** used to show why the evidence supports the claim?

If the answer to any of these questions is no, then the explanation is not valid. You can use these criteria to evaluate and improve the quality of your own explanations and also to evaluate other people's explanations.

In later lessons, you will learn about other parts that are needed for a valid explanation and additional questions you can use to evaluate the quality of explanations.

Pulling It Together

Work on your own to answer these questions. Be prepared for a class discussion.

1. Now that you have learned something about the starting substances and ending substances for the three changes, how would you answer the Key Question, **How do we know if a different substance has been made?**

2. When grayish metal was added to a container of yellow gas, white crystals appeared inside the container. Properties of the starting substances and of the white crystals that form were observed and measured by a scientist and are shown in Table 1.7.

Table 1.7. Gray Metal and Yellow Gas

Appearance	Conductivity	Soluble in Water	Conductive When Dissolved
Metallic, grayish solid	High	No	N/A
Yellow gas	Not conductive	Yes	Yes
White crystals	Low	Yes	Yes

a. Was a different substance made when the grayish metal was mixed with the yellow gas? Use the table below to record notes before writing an explanation. Remember to use the Explanation Quality Questions as a guide.

Question	Was a different substance made when the grayish metal was added to the container of yellow gas?
Claim	
Science Ideas	
Evidence	

Explanation:

Lesson 1.3—Making New Substances From Other Substances

What do we know and what are we trying to find out?

In Lesson 1.2, you investigated three changes in which different substances were made by mixing other substances together. You learned that in order to determine if a new substance forms during a change, you should measure the characteristic properties of the substance.

You may not know it, but changes that produce new and different substances are occurring all around you every day. These changes are even happening in your own body right now, just like they are happening in every living thing. Can you think of any changes in matter that you have observed in your everyday life that might involve making new substances? For example, when you bake a cake, the cake has different properties from the ingredients used to make it. Where does the cake come from? What happens to the original ingredients you mixed to make the cake?

Answer the Key Question to the best of your knowledge. Be prepared to share your ideas with the class.

Key Question: As a new substance forms, what happens to the starting substances?

Activity 1: Observing Three Changes Again

Materials
For each team of students
Vinegar
Baking soda
Teaspoon
Measuring cup
Quart-size zipper plastic bag

In this activity, you will again investigate the three changes from Lesson 1.2. Your teacher will show videos of two changes. You will investigate the other change in teams at your tables. This time as you observe each change, look for changes in the amounts of starting and ending substances to see if you can figure out where the new substances come from.

Procedures and Questions

1. Watch the video of the change involving hexamethylenediamine and adipic acid. Look closely at how the amounts of starting and ending substances are changing from the beginning to the end of the video. Record the names of the substances and your observations in Table 1.8.

Table 1.8. Hexamethylenediamine and Adipic Acid

Describe what happens to the amounts of the starting substances.	
Name	What happens to the amount?
Name	What happens to the amount?
Describe what happens to the amounts of the ending substances.	
Name	What happens to the amount?
Name	What happens to the amount?

2. Use your observations to answer the following questions about the starting and ending substances:

   a. What happens to the amounts of the two liquids as more nylon thread is pulled from the beaker? What is your evidence?

   b. Where do you think the new substance (nylon) is coming from? Support your claim with evidence.

3. Watch the video of the change involving steel wool (iron) and oxygen in the air. Look closely at how the amounts of starting and ending substances are changing from the beginning to the end of the video. Record the names of the substances and your observations in Table 1.9.

Table 1.9. Steel Wool (Iron) and Air (Oxygen)

Describe what happens to the amounts of the starting substances.	
Name	What happens to the amount?
Name	What happens to the amount?
Describe what happens to the amounts of the ending substance.	
Name	What happens to the amount?

4. Use your observations to answer the following questions about the starting and ending substances:

   a. What happens to the amount of oxygen gas in the flask as more rust forms? What is your evidence?

b. What happens to the amount of blackish-gray (unrusted) steel wool in the flask as more rust forms? What is your evidence?

c. Where do you think the new substance (rust) is coming from?

5. Now we will investigate the change involving baking soda and vinegar again.
   a. Carefully place 1 teaspoon of baking soda and the cup containing 25 ml of vinegar in the bag. Be careful not to let the substances mix. Smell the starting substances.
   b. Squeeze as much of the air out of the bag as possible and seal it. Then turn the cup over and mix the baking soda and vinegar thoroughly.
   c. Look closely at how the amounts of starting and ending substances are changing. Record the names of the substances and your observations in Table 1.10.

Table 1.10. Baking Soda and Vinegar

Describe what happens to the amounts of the starting substances.	
Name	What happens to the amount?
Name	What happens to the amount?
Describe what happens to the amounts of the ending substances.	
Name	What happens to the amount?
Name	What happens to the amount?
Name	What happens to the amount?

6. Use your observations to answer the following questions about the starting and ending substances:

   a. What happens to the amount of baking soda in the bag as more carbon dioxide gas forms? What is your evidence?

   b. Open the bag and smell the contents. Based on smell, what do you think happens to the amount of vinegar in the bag as more carbon dioxide gas forms? What is your evidence?

   c. Where do you think the new substance (carbon dioxide) is coming from?

7. Write a general statement that describes how changes in the amounts of the starting and ending substances are related.

8. What does your general statement suggest about where the ending substances come from?

9. If we had the right equipment, we could actually measure the amounts of the starting and ending substances at different time points during the change involving hexamethylenediamine and adipic acid. The first two graphs below show how the amount of hexamethylenediamine and the amount of nylon changed, starting from the time we mixed the starting substances (start) to the time all the starting substances were gone (end). Draw a line on each of the two bottom graphs to show how the amounts of adipic acid and water changed over time.

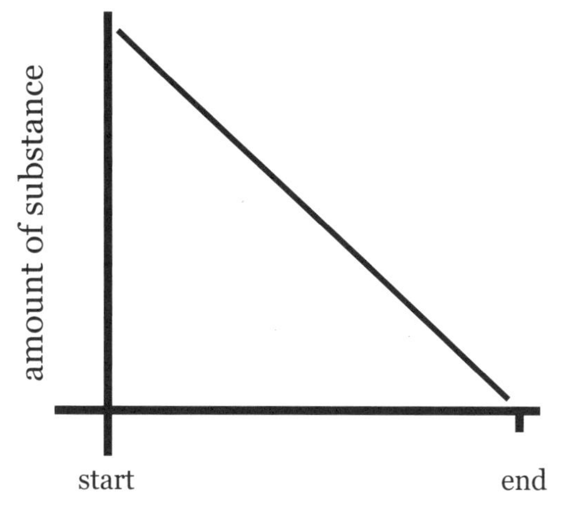

Hexamethylenediamine

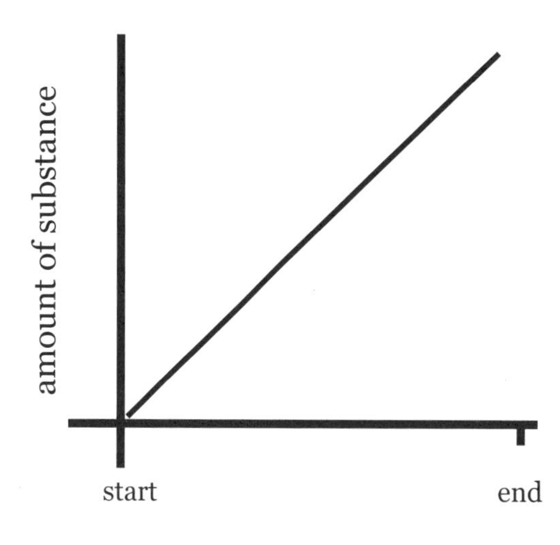

Nylon

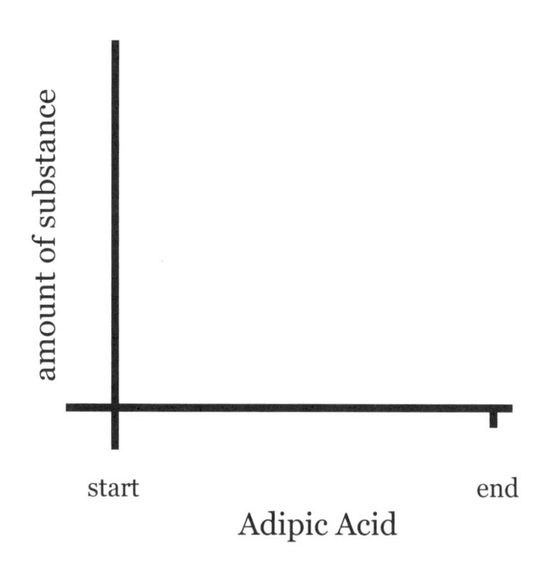

Adipic Acid

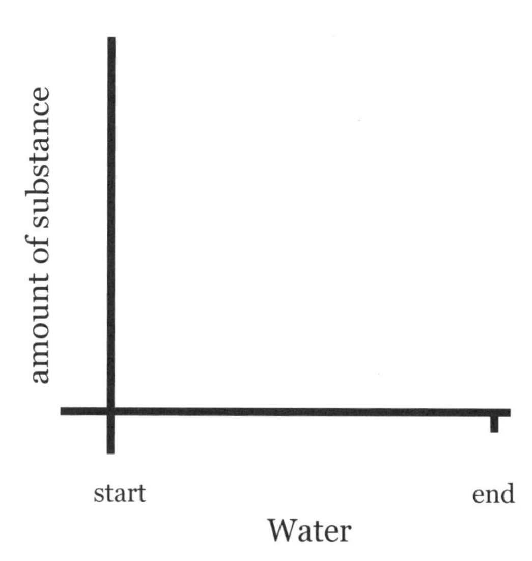

Water

Science Ideas

Science ideas are accepted principles or generalizations about how the world works based on a wide range of observations and data collected and confirmed by scientists. Because these science ideas are consistent with the available evidence, you are justified in applying science ideas to new observations and data.

Lesson 1.3 was intended to help you understand an important idea about how to tell when new substances form and where they might come from. Read the idea below. Look back through Lesson 1.3. In the space provided after the science idea, give evidence that supports the idea.

Science Idea #2: Changes during which new substances form are called chemical reactions. The correlation of increasing amounts of ending substances with decreasing amounts of the starting substances provides evidence that the new substances result from an interaction between the starting substances.

Evidence:

Activity 2: Constructing Scientific Explanations—Part II

In Lesson 1.2, you examined Eva's explanation for her claim that a new substance formed when hexamethylenediamine and adipic acid were mixed and decided that her explanation answered all of the Explanation Quality Questions. In this activity, you will use the Explanation Quality Questions to evaluate an explanation and improve it.

Procedures and Questions

1. Read Jeremy's explanation to the question, Did a new substance form when baking soda and vinegar were mixed, and if so, where did it come from?

   Carbon dioxide was produced from a chemical reaction between baking soda and vinegar. If a new substance is produced, then a chemical reaction has occurred (Science Idea #2) and the properties changed when baking soda and vinegar were mixed. My evidence that carbon dioxide was produced is that it turned limewater cloudy.

2. Now answer the Explanation Quality Questions for Jeremy's explanation. You can refer to Table 1.6 in Lesson 1.2 to refresh your memory about what each part is.

- Does the **claim** respond to the question?

- Are the **science ideas** that are listed relevant to the claim?

- Does the **evidence** support the claim?

- Are the **science ideas** used to show why the evidence supports the claim?

3. Now it is your turn to write an explanation to the question, Did a new substance form when baking soda and vinegar were mixed, and if so, where did it come from?

   a. Record your notes for each part of the explanation in the table below:

Question	Did a new substance form when baking soda and vinegar were mixed, and if so, where did it come from?
Claim	
Science Ideas	
Evidence	

   b. Use the information from the table to write your explanation.

Pulling It Together

Work on your own to answer these questions. Be prepared for a class discussion.

1. After observing the three changes again, how would you answer the Key Question, **As a new substance forms, what happens to the starting substances?**

2. A friend tells you that the burning of a candle involves a chemical reaction, during which oxygen in the air interacts with the candle to produce carbon dioxide and water. Your friend provides evidence listed in the table below. For each statement, indicate what it provides evidence for.

Observation	What It Provides Evidence for
A large beaker is placed over a burning candle. Before the beaker is put over the candle, a burning splint stays lit. However, after the beaker is placed over the candle and the flame on the candle goes out, the air in the beaker puts out the burning splint.	
A large beaker is placed over a burning candle. Before the beaker is put over the candle, the air in the beaker does not turn limewater cloudy. However, after the beaker is placed over the candle and the flame on the candle goes out, the air in the beaker turns limewater cloudy.	
When a large beaker is placed over a burning candle, as the amount of the candle decreases, the amount of oxygen in the beaker also decreases. Meanwhile, the amount of carbon dioxide and water in the beaker increases.	

Lesson 1.4—Using Models to Think About Atoms and Molecules

What do we know and what are we trying to find out?

So far, we've learned that every substance has a unique set of properties that can be observed and measured. We also know that during chemical reactions, substances form that have different properties from the starting substances. How can this happen? If a substance can't change its characteristic properties, where do the new substances come from? Why do different substances have different properties?

To answer these questions, we need to know more about the matter that substances are made of. You may have heard that all matter—that is, every single substance—is made up of atoms. But if all substances are made up of atoms, how can two substances be different? If substances are made up of the same atoms, will they always be the same? Because atoms are too small to be seen, in this lesson you will use models to help you think about these things.

Answer the Key Question to the best of your knowledge. Be prepared to share your ideas with the class.

Key Question: Why do the substances produced in a chemical reaction have different properties from the starting substances?

Activity 1: Comparing Models of Different Substances

Materials
For each team of students
Card Pack 3
Model Key

In this activity, you will look at cards that show different kinds of models of the molecules of different substances. The cards show chemical formulas, structural formulas, LEGO models, ball-and-stick models, and space-filling models, which can all be used to represent the molecules of each substance. You will analyze and interpret these models to help you describe what different substances are made up of and to help you explain why different substances have different properties.

Procedures and Questions

1. Find the Water card in *Card Pack 3* and study the different models on it.

   Use the *Model Key* to help you make sense of the models you are looking at.

2. Answer the questions below based on your observations of the models for a water molecule (see Table 1.11, p. 36).

   a. How does each model represent a water molecule?

   b. How is an atom different from a molecule?

   c. How would you explain to a friend what water is made up of? Be as descriptive as possible.

Table 1.11. Models of a Water Molecule

Model Type	Models of a Water Molecule
LEGO model	Hydrogen Oxygen
Ball-and-stick model	Oxygen Hydrogen
Space-filling model	Oxygen Hydrogen
Chemical formula	$\mathbf{H_2O}$
Structural formula	H–O–H

3. Participate in a class discussion about the similarities and differences between the models and between the models and the real thing that they represent.

4. Spread the rest of the cards from *Card Pack 3* face up on your desk or table.

5. Study the different models on each card. Think about what each substance is made up of:

   a. Does the model show that this substance is an atom or a molecule?

   b. If it is a molecule, then what type(s) of atoms and how many of each make up the molecule?

6. Fill in Table 1.12. What is each substance made up of? Two examples have been done for you.

Table 1.12. What Are Substances Made Up Of?

Substance Name	Observable Properties	Atom or Molecule	Composition
Iron oxide (rust)	Reddish orange flaky solid	Molecule	3 oxygen atoms and 2 iron atoms
Acetic acid (vinegar)			
Iron (steel wool)			
Carbon dioxide gas			
Sodium bicarbonate (baking soda)			
Oxygen gas			
Water	Colorless liquid	Molecule	1 oxygen atom and 2 hydrogen atoms

Activity 2: Comparing Starting and Ending Substances

Materials
For each team of students
Card Pack 1 (green outline)
Card Pack 2 (yellow outline)
Card Pack 3
Model Key
Blank Change Charts

In this activity, you will learn more about the atoms and molecules that make up the starting and ending substances involved in the three chemical reactions that we have been exploring. You will use information on substance cards to think about why the new substances formed during a chemical reaction have properties that are different from those of the starting substances.

Procedures and Questions

1. Find the *Change Chart* for Steel Wool (Iron) and Air (Oxygen). Look in *Card Pack 1* for the two starting substance cards, and place them on the left (starting substance) side of the chart. Then look in *Card Pack 2* for the ending substance, and place that card on the right (ending substance). You already know that the ending substance (iron oxide) has different properties from the starting substances (iron and oxygen). Let's see if we can use the models of the substances in *Card Pack 3* to figure out why.

2. First, using *Card Pack 3*, look at the types of atoms making up the molecules of the starting substances and the ending substances, and record the information in the first row of Table 1.13.

3. Look at the ball-and-stick models or the structural formulas for the starting and ending substances, and describe how the atoms are arranged in the second row of Table 1.13.

Table 1.13. Steel Wool (Iron) and Air (Oxygen)

	Starting Substances	Ending Substances
Types of Atoms		
Arrangement of Atoms		

4. How do the types of atoms in the starting substances compare with the types of atoms in the ending substances?

5. Compare the arrangement of the atoms in the starting and ending substances.

6. Based on the information in Table 1.13, why do you think iron oxide might have different properties from iron and oxygen?

7. Find the *Change Chart* for Hexamethylenediamine and Adipic Acid. Look in *Card Pack 1* for the two starting substance cards, and place them on the left (starting substance) side of the chart. Then look in *Card Pack 2* for the ending substances, and place those cards on the right (ending substance).

8. Then, using *Card Pack 3,* look at the types of atoms making up the molecules of the starting substances and the ending substances, and record the information in the first row of Table 1.14.

9. Look at the ball-and-stick models or the structural formulas for the starting and ending substances, and describe how the atoms are arranged in the second row of Table 1.14.

Table 1.14. Hexamethylenediamine and Adipic Acid

	Starting Substances	**Ending Substances**
Types of Atoms		
Arrangement of Atoms		

10. How do the types of atoms in the starting substances compare with the types of atoms in the ending substances?

11. Compare the arrangement of the atoms in the starting and ending substances.

12. Based on the information in Table 1.14, why do you think nylon and water might have different properties from heyxamethylenediamene and adipic acid?

13. Find the *Change Chart* for Baking Soda and Vinegar. Look in *Card Pack 1* for the two starting substance cards, and place them on the left (starting substance) side of the chart. Then look in *Card Pack 2* for the ending substances, and place those cards on the right (ending substance).

14. Look at the types of atoms making up the molecules of the starting substances and the ending substances, and record the information in the first row of Table 1.15.

15. Look at the ball-and-stick models or the structural formulas for the starting and ending substances, and describe how the atoms are arranged in the second row of Table 1.15.

Table 1.15. Baking Soda and Vinegar

	Starting Substances	**Ending Substances**
Types of Atoms		
Arrangement of Atoms		

16. How do the types of atoms in the starting substances compare with the types of atoms in the ending substances?

17. Compare the arrangement of the atoms in the starting and ending substances.

18. Based in the information in Table 1.15, why do you think sodium acetate, water, and carbon dioxide might have different properties from baking soda and vinegar?

19. Based on what you have written for each of the three reactions, write a general statement that explains why the ending substances of a chemical reaction have different properties from the starting substances.

Science Ideas

Activities 1 and 2 were intended to help you understand what substances are made of and give you ideas about why different substances have different properties. Read the science ideas listed below. Remember that because science ideas are consistent with a wide range of relevant evidence, you can use them to explain observations and data about other substances. Look back through Lesson 1.4 and give at least one piece of evidence that supports each science idea.

Science Idea #3: A molecule is made up of two or more atoms connected together in a specific arrangement.

Evidence:

Science Idea #4: Each substance is made up of a single type of atom or molecule. The properties of a substance are determined by the type, number, and arrangement of atoms that it is made up of. Because no two substances are made up of the same arrangement of atoms, no two substances have the same set of properties.

Evidence:

Pulling It Together

Work on your own to answer these questions. Be prepared for a class discussion.

1. After studying the models, how would you answer the Key Question, **Why do the substances produced in a chemical reaction have different properties from the starting substances?**

2. Recall the scientist who collected the white crystals that resulted from mixing a yellow gas and grayish metal together. She determined that the chemical formula for the crystals was NaCl. She then took the white crystals from the container and ground them down into a very fine powder. The texture of the powder was different from the texture of the crystals. But when the scientist measured the conductivity and solubility of the powder, she found them to be the same as the conductivity and solubility of the crystals. She then determined that the chemical formula for the powder was NaCl.

    a. Is a new substance made when the crystals are ground down into a powder? Use the table below to make notes and then write a valid explanation.

Question	Is a new substance made when the crystals are ground into a powder?
Claim	
Science Ideas	
Evidence	

Explanation:

3. In this lesson, you've seen how different kinds of models can be used to show the atoms and molecules that make up the starting and ending substances involved in chemical reactions. You've also used the models to think about why the starting and ending substances have different properties, even though they are both made of the same types of atoms. How do you think you can use these same models to help you figure out why new substances are produced in chemical reactions?

Lesson 1.5—Using Models to Represent Chemical Reactions

What do we know and what are we trying to find out?

An important idea from our previous lessons was that different substances have different properties. This is because every substance is made up of its own unique combination and arrangement of atoms, which gives each substance its own unique set of properties.

So far, we have been investigating three chemical reactions:

- Steel Wool (Iron) and Air (Oxygen)
- Hexamethylenediamine and Adipic Acid
- Baking Soda and Vinegar

As we have learned, during chemical reactions, one or more substances "react" with each other and form one or more different substances. How does this happen?

In Lessons 1.5 and 1.6, you will explore what happens to the atoms that make up the molecules of the starting substances when new substances are formed during chemical reactions. You will use models of the molecules to help guide your thinking.

Answer the Key Question to the best of your knowledge. Be prepared to share your ideas with the class.

Key Question: How are the molecules of the ending substances made?

Activity 1: Representing Chemical Reactions With Models

Materials
For each team of students
Chemical Reaction Mats: Steel Wool (Iron) and Air (Oxygen), Baking Soda and Vinegar
LEGO model kit
Model Key

In this activity, you will use LEGO bricks to represent (or to model) what is happening to the atoms and molecules in two of the three chemical reactions we have been studying. You will use the *Chemical Reaction Mats* to guide you.

The *Chemical Reaction Mats* show the LEGO models of starting substances on one side (green side), and LEGO models of the ending substances on the other (yellow side). Each LEGO brick represents an atom. You can use the information on each mat to write a word equation for that chemical reaction. The general form of a word equation is shown below:

Starting substances ⟶ **Ending substances**

The arrow means "react to form." When you flip from the green side to the yellow side of the *Chemical Reaction Mat*, you are representing "react to form."

Procedures and Questions

1. Using the *Steel Wool (Iron) and Air (Oxygen) Chemical Reaction Mat* and LEGO model kit, do the following:
   a. Use LEGO bricks to build models of the molecules of the starting substances shown on the green side of the mat.
   b. Once you have built the models of the starting molecules exactly as they are shown on the mat, put all unused LEGO bricks back in the model kit and close the kit.
   c. Look closely at the starting molecule models you built and the ending molecule models shown on the yellow side of the mat. Think about how the models of starting molecules and ending molecules are similar to and different from each other.
   d. Using only the LEGO bricks of the starting molecule models, build models of the molecules of the ending substances as shown on the yellow side of the mat. Use the *Model Key* to help you make sense of what you are looking at.

2. Answer these questions based on your observations:

   a. What is represented by the *Chemical Reaction Mat?* Use the information on the *Chemical Reaction Mat* to write a word equation for the chemical reaction using the following structure:

   Starting substances ⟶ **Ending substances**

   b. When you built the models of the ending molecules from models of the starting molecules, did you have any leftover LEGO bricks (atoms)?

   c. Did you need any additional LEGO bricks (atoms)?

3. Disassemble your LEGO models.
4. Repeat Step 1, using the *Baking Soda and Vinegar Chemical Reaction Mat.*
5. Answer these questions based on your observations:

   a. What is represented by the *Chemical Reaction Mat?* Use the information on the *Chemical Reaction Mat* to write a word equation for the chemical reaction using the following structure:

   Starting substances ⟶ **Ending substances**

   b. When you built the models of the ending molecules from models of the starting molecules, did you have any leftover LEGO bricks (atoms)?

   c. Did you need any additional LEGO bricks (atoms)?

d. How would you describe what you did to the LEGO bricks (atoms) in order to make the models of the ending molecules from the models of the starting molecules?

e. Did you always have to disconnect every LEGO brick (atom) of the models of the starting molecules to make the models of ending molecules? Give a specific example from the two reactions.

6. Based on what you learned with the models, write a general rule or statement that explains what happens to the atoms in a chemical reaction.

7. Apply your thinking to predict which ending substances could form when methane gas reacts with oxygen gas from the air. Table 1.16 shows space-filling models for these two starting substances. For each substance shown in Table 1.17, indicate whether it is or is not a possible ending substance and explain why.

Table 1.16. Starting Substances

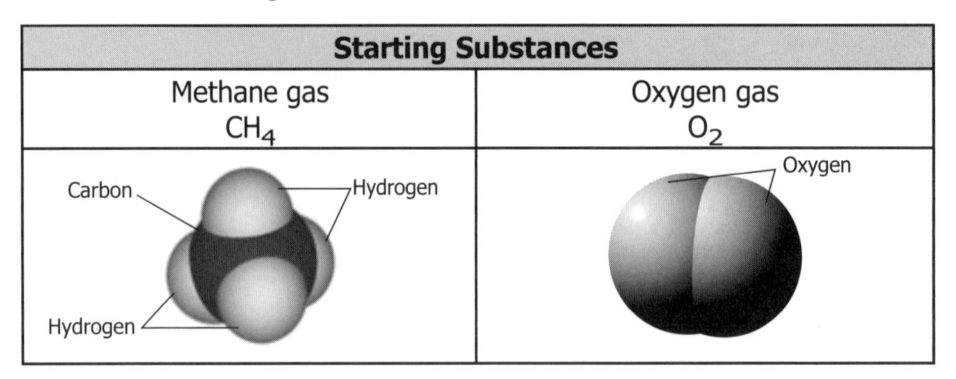

Starting Substances	
Methane gas CH_4	Oxygen gas O_2
Carbon Hydrogen Hydrogen	Oxygen

Table 1.17. Possible Ending Substances

Possible Ending Substances	Fill in the blank in each sentence with "is" or "is not," and write why you think this.
Water H_2O Oxygen Hydrogen	I think water ______ a possible ending substance because ...
Carbon dioxide CO_2 Oxygen Carbon	I think carbon dioxide _______ a possible ending substance because ...
Carbon monoxide CO Oxygen Carbon	I think carbon monoxide _______ a possible ending substance because ...
Ammonia NH_3 Nitrogen Hydrogen	I think ammonia ______ a possible ending substance because ...

Science Ideas

The activity in this lesson was intended to help you construct an important idea about chemical reactions. Read the idea below. Notice that the second sentence of Science Idea #5 explains the formation of new substances during chemical reactions in terms of atom rearrangement. In this unit, we will often explain things that we can see, such as the formation of substances like rust or water, in terms of things we can't see—namely, what is happening to the atoms and molecules. We can observe firsthand or infer from data when new substances are produced, but we can't see the atoms rearranging. However, because Science Idea #5 states a general principle that is consistent with a wide range of observations and data, we can use it to reason that whenever we see that one or more new substances are produced, we know that atoms must have rearranged. **From now on, you will be expected to use ideas about atom rearrangement whenever you are asked to explain why one or more new substances form.**

Look back through Lesson 1.5. In the space provided after the science idea, give at least one piece of evidence that supports the idea.

Science Idea #5: During chemical reactions, atoms that make up molecules of the starting substances (called reactants) disconnect from one another and connect in different ways to form the molecules of the ending substances (called products). Because the arrangement of atoms in the products is different from the arrangement of atoms in the reactants, the products of a chemical reaction have different properties from the reactants.

Evidence:

Activity 2: Constructing Scientific Explanations—Part III

Just as scientists often do, you have used models of chemical reactions to visualize a process that takes place at a scale too small to be seen. The models you have used to help visualize and think about atom rearrangement are based on a great deal of scientific evidence (see Table 1.18). In this activity, you will use what you have learned from the models to help explain what might be happening during chemical reactions.

Table 1.18. Parts of an Explanation and Criteria for Evaluating Its Quality

	What It Is	Explanation Quality Criteria
Question	The question to be answered	
Claim	A statement or conclusion that is intended to respond to the question	• The claim responds to the question.
Science Ideas	Widely accepted scientific ideas, concepts, or principles that can be used to show why the evidence supports the claim	• The science ideas that are listed (or paraphrased) are relevant to the claim. • The science ideas are used to show why the evidence supports the claim. (This involves applying the science ideas to the specific case in the question.)
Evidence	Data that are relevant to and support the claim and can be confirmed by others Data are based on observations about the world that are either made with our senses or measured with instruments.	• The evidence supports the claim. (All the data should be consistent with the science idea, not just the data that are cited as evidence.)
Models	Tools for thinking about the world that can give you ideas about how something might work	• The models are consistent with the science ideas listed. • The models show that the claim is reasonable.

Remember that during Lesson 1.3 we answered the question, Did a new substance form when baking soda and vinegar were mixed, and, if so, where did it come from? Now we have added *models* as another part of a valid explanation. We have also added two new Explanation Quality Criteria to help us think about models. Let's see how one student used models to reason about atoms and molecules in his claim.

1. Examine the student's notes and how they were put together to construct a valid explanation.

Question	Did a new substance form when baking soda and vinegar were mixed, and if so, where did it come from?
Claim	*Yes, new substances formed that came from a chemical reaction between baking soda and vinegar. I think the Na, H, C, and O atoms making up carbon dioxide, water, and sodium acetate came from the baking soda and vinegar. These atoms originally made up the baking soda and vinegar and must have disconnected and then connected in new ways to form the molecules of the new substances.*
Science Ideas	*Changes during which new substances form are called chemical reactions. The correlation of increasing amounts of ending substances with decreasing amounts of the starting substances provides evidence that the new substances result from an interaction between the starting substances (Science Idea #2).* *New substances form during chemical reactions because atoms that make up molecules of the reactants rearrange to form the molecules of the products, which have different properties from the reactants (Science Idea #5).*
Evidence	*Evidence a new substance was produced: The ending substances have different properties from the starting substances. Baking soda is a white solid, and vinegar is a clear, colorless, smelly liquid. When they were mixed, a colorless, odorless gas was produced that gave a positive limewater test.* *Evidence the carbon dioxide came from the chemical reaction between baking soda and the vinegar: We observed that after baking soda and vinegar were mixed, the smelly liquid and the white powder were gone but a gas that gave a positive limewater test appeared. This test gives evidence that carbon dioxide is present.* *Evidence sodium acetate ($NaC_2H_3O_2$) is also produced: Reactants baking soda ($NaHCO_3$) and vinegar ($HC_2H_3O_2$) have Na, H, C, and O atoms; but carbon dioxide (CO_2) doesn't have any Na or H atoms. The formula on the substance card for sodium acetate ($NaC_2H_3O_2$) shows that it has Na and H atoms.* *Evidence water (H_2O) is also produced: At the end of the chemical reaction, a clear, colorless liquid is present in the plastic bag. However, CO_2 is a gas and $NaC_2H_3O_2$ is a white solid. Water is a clear colorless liquid, and sodium acetate dissolves in water.*

Models	Using LEGO models of the molecules of baking soda and vinegar, we showed it was possible to rearrange the LEGO bricks to form molecules of carbon dioxide, water, and sodium acetate. So the Na, H, C, and O atoms making up baking soda and vinegar must have disconnected and connected in new ways to form the molecules of carbon dioxide, sodium acetate, and water.

Explanation: New substances formed from a chemical reaction between baking soda and vinegar. Changes during which starting substances interact to form new substances are called chemical reactions (Science Idea #2). The ending substances can be recognized as new substances because they have different properties from the starting substances (Science Idea #5). Baking soda is a white solid, and vinegar is a clear, colorless, smelly liquid. When they were mixed, a gas appeared that wasn't initially present. The gas gave a positive limewater test. Because a positive limewater test indicates the presence of carbon dioxide, we have evidence that the gas is carbon dioxide. Because carbon dioxide has different properties from vinegar and baking soda, we have evidence of a chemical reaction.

The atoms that make up the new substances (Na, H, C, and O) come from the molecules that made up the baking soda ($NaHCO_3$) and vinegar ($HC_2H_3O_2$). New substances form during chemical reactions because atoms that make up the molecules of the reactants disconnect and reconnect to form the molecules of the products (Science Idea #5). This means that the types of atoms in the products of a chemical reaction are the same as the types of atoms in the reactants. The substance cards of the reactants, which give the chemical formulas of the molecules, show that baking soda ($NaHCO_3$) is made up of Na, H, C, and O atoms, and vinegar ($HC_2H_3O_2$) is made up of H, C, and O atoms. Because both baking soda and vinegar are made up of C and O atoms, the C and O atoms of the CO_2 could have come from either the baking soda or the vinegar. However, since the product (CO_2) doesn't have any Na or H atoms, there must be one or more products whose molecules contain those atoms. The substance card for sodium acetate ($NaC_2H_3O_2$) shows that it has Na and H atoms, and therefore sodium acetate ($NaC_2H_3O_2$) could be another product.

However, we have evidence for yet another product. There is a clear colorless liquid present at the end of the chemical reaction, and the substance cards show that carbon dioxide is a gas and sodium acetate is a white solid. Since neither carbon dioxide nor sodium acetate is a clear, colorless liquid, there must be another product. Water is a clear, colorless liquid, so water could be another product.

We don't see a white solid at the end of the reaction. However, we observed before that sodium acetate dissolves in water. So maybe water is also a product and the clear, colorless liquid is actually sodium acetate dissolved in water. To confirm, we would need to separate the substances and determine their properties.

2. With your team, evaluate the student's explanation. Discuss the Explanation Quality Criteria listed in Table 1.18. Underline the part of the student's explanation where models were used to help in the student's reasoning.

3. Now write an explanation to the question, Did a new substance form when steel wool was exposed to oxygen, and if so, where did the atoms that make up the new substance come from?

Use the table to help you organize your thoughts. Remember to reason using atoms and molecules in your explanation.

Question	Did a new substance form when steel wool was exposed to oxygen, and if so, where did the atoms that make up the new substance come from?
Claim	
Science Ideas	
Evidence	

Models	

Explanation:

4. Discuss how your explanation meets the Explanation Quality Criteria.

Pulling It Together

Work on your own to answer these questions. Be prepared for a class discussion.

1. Summarize your learning in this lesson by answering the Key Question, **How are the molecules of the ending substances made?** Using atoms and molecules, explain your answer in a way that you would to a friend who missed class today.

2. In this unit, you have also seen the chemical reaction between hexamethylenediamine and adipic acid. The chemical formula for hexamethylenediamine is $C_6H_{16}N_2$, and the chemical formula for adipic acid is $C_6H_{10}O_4$. Based on what you now know about atom rearrangement in chemical reactions, what atoms do you predict will be present in the products (ending substances)? Support your answer.

3. Hydrogen peroxide is a clear, colorless liquid. When you apply it to a cut or scrape on your skin, you will see bubbles and you may hear a "fizzing" sound. Figure 1.3 uses LEGO models to represent hydrogen peroxide on the left. On the right are models showing what happens to hydrogen peroxide when you apply it to a cut or scrape on your skin. Your teacher will project a color version of Figure 1.3 so that you can see the models more clearly.

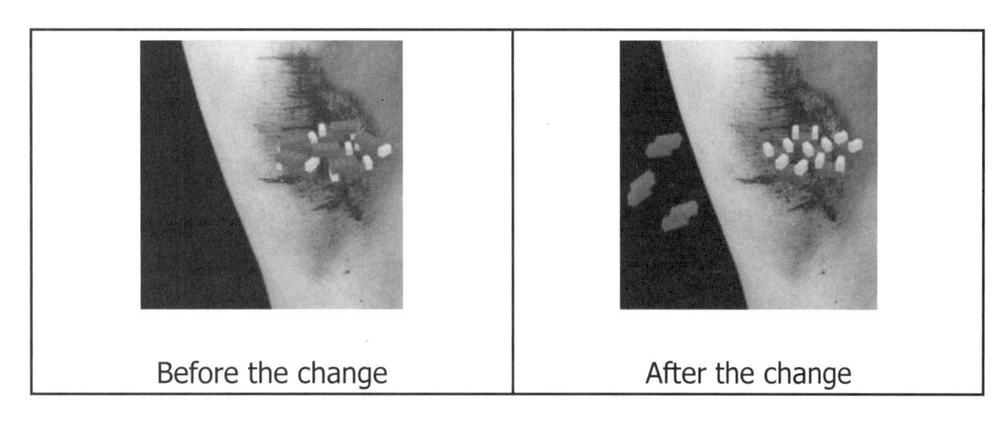

Figure 1.3. Hydrogen Peroxide Before (left) and After (right) Change

Do you think a chemical reaction occurs when hydrogen peroxide is put on a wound? Explain why you do or do not think so in terms of atoms and molecules. Your explanation should use science ideas, evidence, and models to support your claim. (Use the table for your notes.)

Question	Is the change hydrogen peroxide undergoes an example of a chemical reaction?
Claim	
Science Ideas	
Evidence	

Models	

Explanation:

a. Underline the part of your explanation where you used models in your reasoning.

b. How does your explanation meet the Explanation Quality Criteria? Give details. (See Table 1.18 on p. 51 for a reminder of what the criteria are.)

4. As a pot of water is heated on a stove, bubbles of gas form in the liquid. Figure 1.4 uses LEGO models to represent the molecules of water in the pot on the left. On the right are models of the molecules inside the bubbles that form.

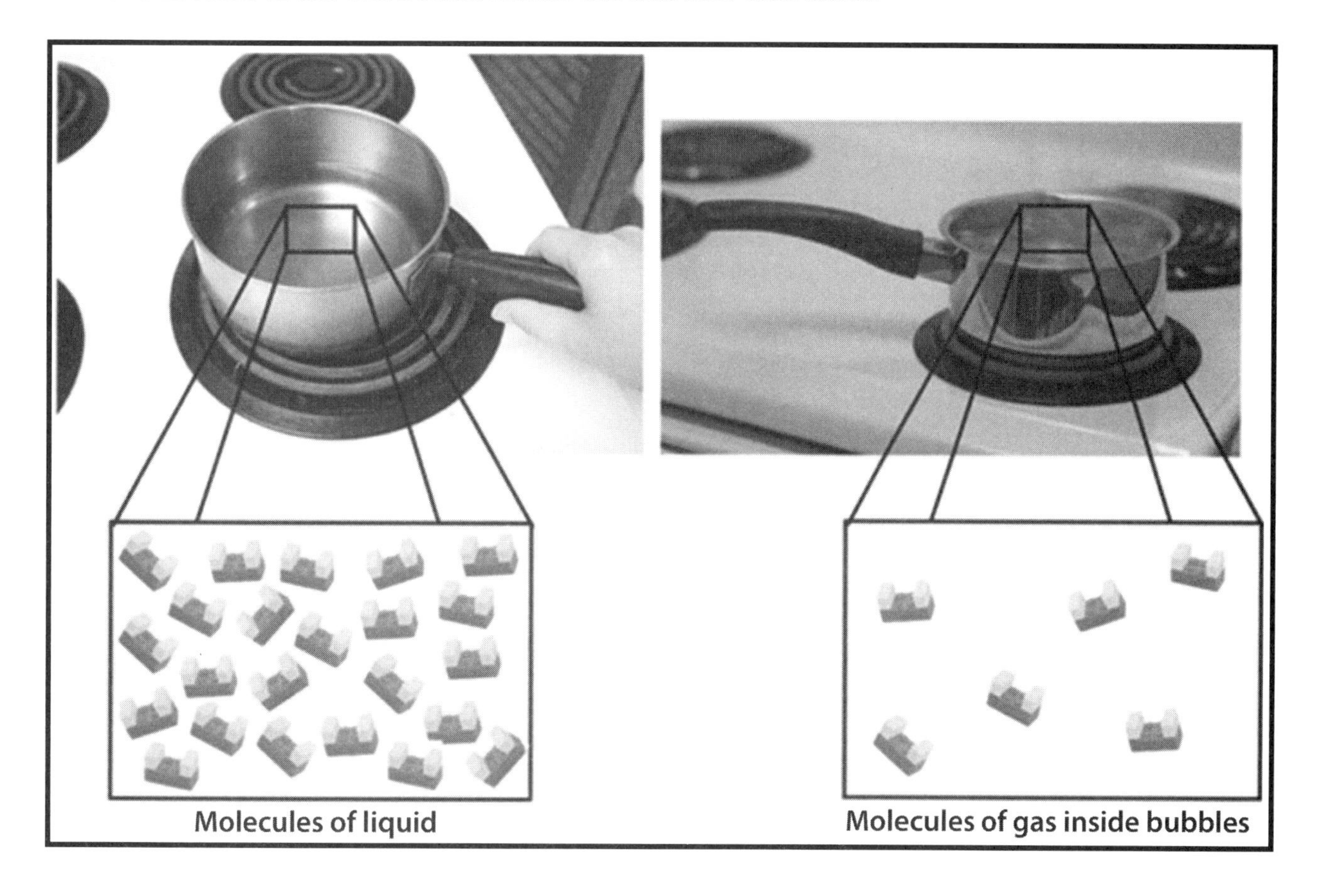

Figure 1.4. Water Being Heated on Stove

Do you think a chemical reaction occurs when water is heated on a stove? Explain why you do or do not think so, using science ideas, evidence, and models to support your claim. (Use the table for your notes.)

Question	Does a chemical reaction occur when water is heated on a stove?
Claim	
Science Ideas	

Evidence	
Models	

Explanation:

Lesson 1.6—Representing Chemical Reactions That Produce Large Molecules

What do we know and what are we trying to find out?

You have used LEGO bricks to represent (or model) what happens when atoms from small molecules rearrange to form other small molecules. But some kinds of molecules are very large (made up of thousands of atoms!). Many human-made materials (for example, nylon, rubber, and plastics) that we see and use each day are made up of very large molecules.

You may not know this, but your body structures, such as muscles and skin, are made up mostly of very large molecules. Plant body structures, such as stems and roots, are also made up of very large molecules. Understanding how very large molecules form from small molecules will eventually help us explain how a puppy and sequoia tree grow, how a wound heals, and how a lizard regrows its tail.

Use what you know about chemical reactions to answer the Key Question.

> **Key Question: How can small molecules be used to make very large molecules?**

Activity 1: Representing Nylon Formation With Models

Materials
For each team of students
Chemical Reaction Mat: Hexamethylenediamine and Adipic Acid
HGS ball-and-stick model kit
Wet or dry erase markers

In this lesson, you will model the chemical reaction that produces nylon. Pay careful attention to what and how new substances form during the reaction. You will use ball-and-stick models that are more like the ones that scientists use than the models you have built with LEGO bricks in other lessons. These ball-and-stick models represent atoms using balls of the same color as the LEGO bricks, but they use sticks to connect atoms. Ball-and-stick models make it easier to represent large molecules because the ball-and-stick models do not fall apart as easily as LEGO models. We will use ball-and-stick models any time we want to represent molecules that are made up of chains of carbon atoms, such as hexamethylenediamine and adipic acid.

Procedures and Questions

1. Working with a partner, use the *Hexamethylenediamine and Adipic Acid Chemical Reaction Mat* and HGS ball-and-stick model kit to complete the steps below.
   a. Use ball-and-stick model pieces to build models of the molecules of the starting substances shown on the green side of the mat. For each molecule, do the following:
      i. Attach the six carbon atoms in a chain first.
      ii. Add the oxygen or nitrogen atoms to the carbon atoms at either end of the carbon atom chain.
      iii. Then add hydrogen atoms to the carbon, oxygen, and nitrogen atoms as shown on the mat. (The orientations of atoms—pointing toward you or away from you—are not important.)
   b. Once you have built the models of the starting molecules exactly as they are shown on the mat, put all unused ball-and-stick model pieces back in the model kit and close the kit.
   c. Turn the mat over to the yellow side, which shows you how to build your models of the ending molecules. Before you begin building, look carefully at the ending molecules shown on the mat so that you can accomplish the following tasks:
      i. Identify anywhere you are going to break a connection. Use a wet or dry erase marker to draw an *X* on the mat over any connections you are going to break.

         Lay a pencil or pen across these same connections on the ball-and-stick models you built to show which ones will break.

ii. Identify where you are going to form any new connections. Use the marker to circle atoms and draw lines on the mat showing any new connections you are going to make. Identify the atoms in the ball-and-stick models that you built to show how atoms will connect.

iii. Have your teacher check your mat and give the OK to make these changes to the models.

d. Use only the ball-and-stick model pieces of the starting molecule models to build models of the ending molecules as shown on the yellow side of the mat.

2. Answer these questions based on your observations:

a. What is represented on the *Chemical Reaction Mat?* Use the information on the *Chemical Reaction Mat* to write a word equation for the chemical reaction.

Starting substances ⟶ **Ending substances**

b. When you built the models of the ending molecules from models of the starting molecules, did you have any leftover balls (atoms)?

c. Did you need any additional atoms?

d. Did you have to disconnect every atom of the models of the starting molecules to make the models of the ending molecules?

e. During this chemical reaction, when you connected the two different small molecules together, what other molecule did you form?

f. Where did the H and OH atoms come from that linked together to form the water molecule? Be as specific as you can.

g. What atoms from each small molecule link together? Be as specific as you can.

You just modeled the basic chemical reaction that happens between two small molecules. But the small molecules can react in two places, not just on one end as you modeled. So if you had many more of the small molecules, you could link them end to end to form a large molecule. We call the large molecules *polymers* (*poly* means many and *mer* means part) because they are made up of many small molecules called *monomers* (*mono* means one and *mer* means part).

The image in Figure 1.5 shows a longer segment of nylon polymer in which each of the monomers at the end has reacted with another monomer. Color the O atoms red and the N atoms blue.

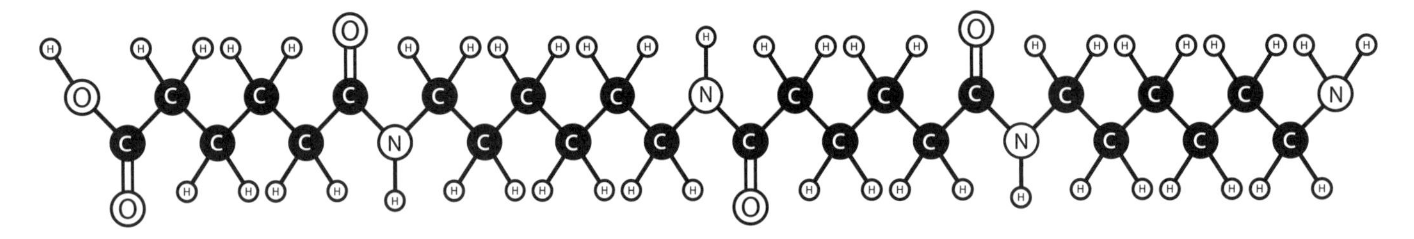

Figure 1.5. Segment of a Nylon Polymer

3. Examine the segment of nylon polymer in Figure 1.5 and answer these questions with your team.

   a. How many adipic acid monomers were used to make this polymer chain? Draw circles around the parts that come from adipic acid.

   b. How many hexamethylenediamine monomers were used to make this polymer chain? Draw rectangles around the parts that come from hexamethylenediamine.

   c. How many water molecules were made?

4. Imagine that we connected all of the class's monomers together one at a time to form a single polymer chain with alternating monomers. Answer these questions about the imaginary polymer:
   a. How many total monomer models did the class make?

   b. How many of each type of monomer were made?

   c. How many water models would be made by the class if all of the monomer models were linked together in a chain?

   d. Write a word equation for the formation of this chain.

   Starting substances ⟶ **Ending substances**

5. Take apart and put away your models.

If a 1,000-mer were made of real atoms instead of ball-and-stick models, it would still be too small for you to see with even the highest-powered microscope. Polymers are extremely large molecules compared with other molecules that you have studied, like water. However, it still takes extremely large numbers of polymer molecules to make up visible amounts of these substances. For example, nylon thread is actually many long polymers twisted together. Just 1 millimeter of nylon thread contains about 100,000,000,000,000,000 monomer molecules. If that many real monomers were represented with ball-and-stick models, the models would fill up more than 50 billion (50,000,000,000) classrooms!

Science Ideas

Lesson 1.6 was intended to expand your understanding of atom rearrangement to include polymer formation. Read the idea below. Remember that because science ideas are consistent with a wide range of relevant evidence, you are justified in applying them to new observations and data. Look back through Lesson 1.6. In the space provided after the science idea, give at least one piece of evidence that supports the science idea.

Science Idea #6: Very large molecules called polymers can be formed by reacting small molecules (monomers) together. Because monomers can react in two places, it is possible for each monomer to react with two other monomers to form long polymer chains. Each time a monomer is added to the chain, atoms are rearranged and another molecule, typically water, is formed.

Evidence:

Pulling It Together

Work on your own to answer the following questions. Be prepared for a class discussion.

1. Answer the Key Question: **How can small molecules be used to make very large molecules?**

2. Figures 1.6 and 1.7 show segments of polymer molecules of cotton and spider silk. Color the O atoms red and the N atoms blue. Look for repeating patterns in each polymer molecule. Draw brackets around the monomers for each polymer molecule.

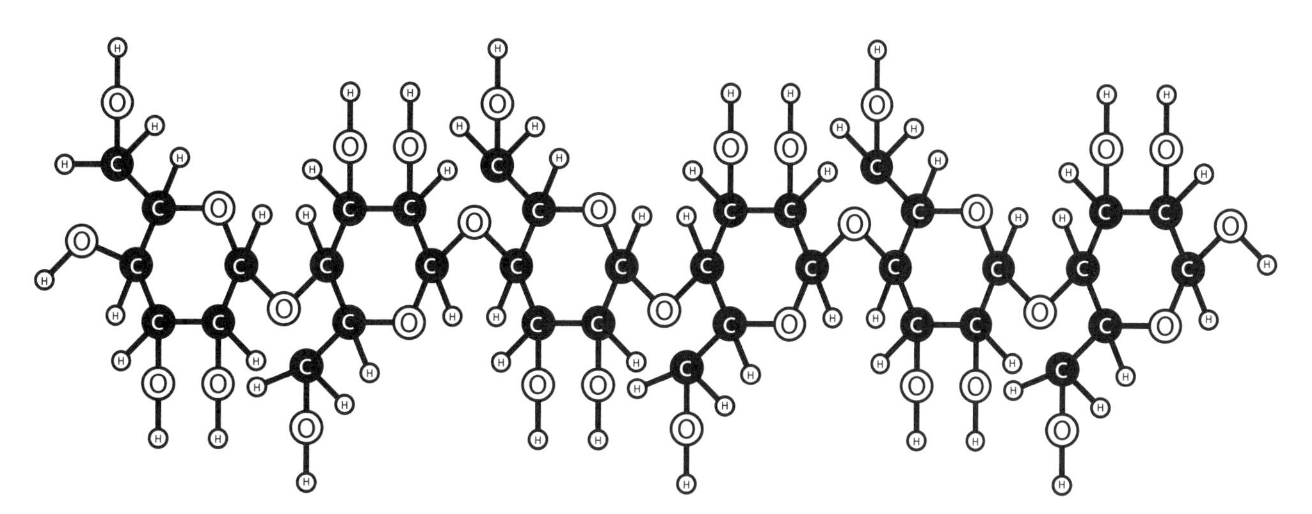

Figure 1.6. Segment of a Cotton Polymer Molecule

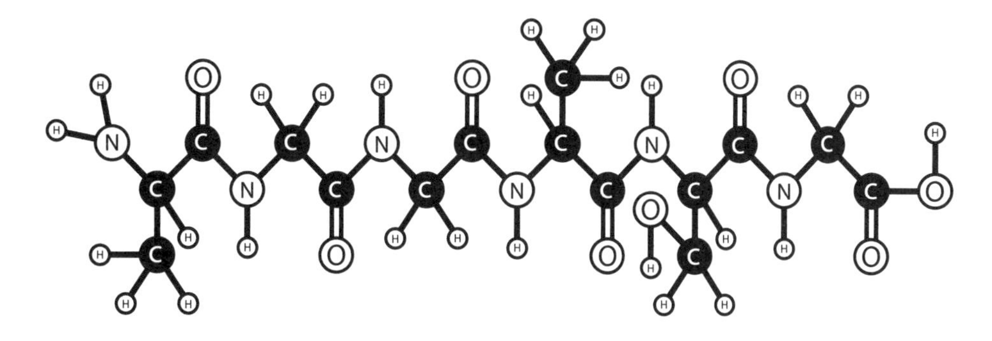

Figure 1.7. Segment of a Spider Silk Polymer Molecule

3. Kevlar is a synthetic polymer that is much stronger than nylon. In fact, the same weight of Kevlar is five times stronger than steel. It is used in bicycle tires, body armor, and drumheads because of this amazing property. Figure 1.8 shows a model of a segment of Kevlar. Color the O atoms red, the N atoms blue, and the Cl atoms green.

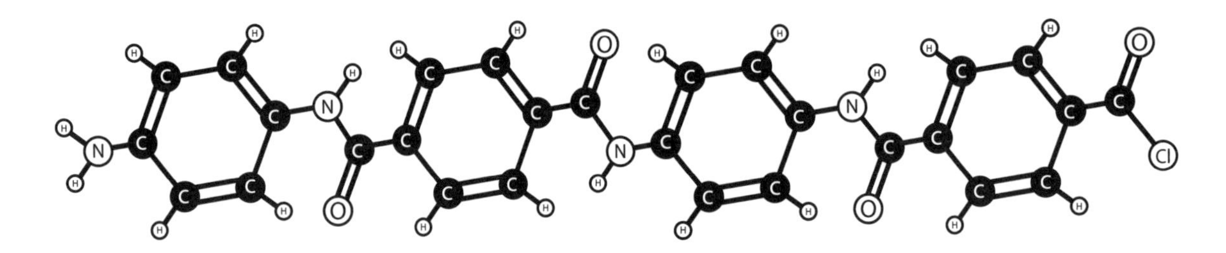

Figure 1.8. Segment of Kevlar Polymer

Figure 1.9 shows the two monomers (para-phenylenediamine and terephthaloyl chloride) that react to form the Kevlar polymer. Color the O atoms red, the N atoms blue, and the Cl atoms green.

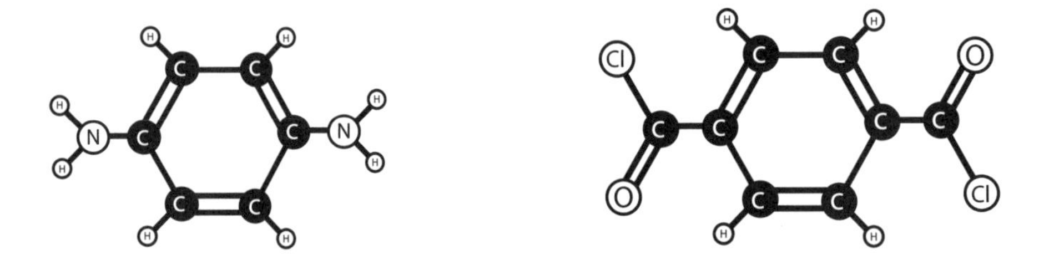

para-phenylenediamine monomer

terephthaloyl chloride monomer

Figure 1.9. Monomers Used to Make Kevlar

a. Circle the atoms of the monomers that rearrange to form the Kevlar polymer.

b. Draw models to represent any other products that you predict will form.

Toward High School Biology

Chapter 2

Student Edition

Lesson 2.1—Chemical Reactions and Mass

What do we know and what are we trying to find out?

When living things grow, they gain weight. For example, a baby weighs about 8 pounds when born. After 6 months, she weighs about twice that—16 pounds. By age 3, she is about 30 pounds. Where does the extra matter come from that makes her weigh more?

In Chapter 1, Lesson 1.5, we saw that new substances are made when atoms that make up the molecules of the starting substances (or reactants) rearrange to form the molecules of the ending substances (or products). Understanding how new substances are formed in simple systems will help us build an explanation of how living things grow and why they gain weight when they grow.

First, let's see if the amount of matter (we call this mass) changes during the chemical reactions we have investigated. To figure this out, we have to keep track of all the matter involved in each chemical reaction. One way to do this is to carry out the reaction in a sealed container so that nothing can enter or leave the container. Then we can open the container and observe if the mass has changed. Later, we will consider what this has to do with the growth of living organisms.

With a partner, think about what happens to the amount of matter during each chemical reaction you have studied so far. Answer the Key Question for one of the chemical reactions listed.

- Steel Wool (Iron) and Air (Oxygen)
- Hexamethylenediamine and Adipic Acid
- Baking Soda and Vinegar

Key Question: What happens to the amount of matter (mass) when new substances form during chemical reactions?

Activity 1: Measuring Mass

Your teacher will demonstrate or show a video of the chemical reaction Baking Soda and Vinegar. Then you will watch a video of the Steel Wool (Iron) and Air (Oxygen) reaction. Each reaction will be carried out in a sealed container. A balance will be used to measure the mass before the chemical reaction. Then the container will be opened, and mass will be measured again.

Procedures and Questions

1. Watch the demonstration or video for the Baking Soda and Vinegar chemical reaction, and record your observations in Table 2.1.

Table 2.1. Mass of the Baking Soda and Vinegar Reaction

Initial mass (g) of the reactants in the sealed container	g
Final mass (g) after the reaction occurs in the sealed container	g
Final mass (g) after opening the container	g
What else do you observe (if anything) when the container is opened?	

2. Answer the questions below based on your observations.
   a. Was the mass that we measured on the scale (measured mass) after the chemical reaction occurred in the sealed container different from the initial mass of the reactants? If so, how was it different?

   b. Did the measured mass change after the container was opened? If so, how did it change?

   c. In previous lessons, you learned that carbon dioxide gas was formed in this reaction. Does carbon dioxide gas have mass? What is your evidence?

d. What do you think happened to cause the measured mass on the scale to change after the container was opened?

e. Draw and label a diagram that shows what you think happened to the amount of matter when the container was opened.

3. Watch the demonstration or video for the Steel Wool (Iron) and Air (Oxygen) chemical reaction, and record your observations in Table 2.2.

Table 2.2. Mass of the Steel Wool (Iron) and Air (Oxygen) Reaction

Initial mass (g) of reactants in the sealed container	g
Final mass (g) after the reaction occurs in the sealed container	g
Final mass (g) after opening the container	g
What else do you observe (if anything) when the container is opened?	

4. Answer the questions below based on your observations.

a. Was the measured mass after the chemical reaction occurred in the sealed container different from the initial mass?

b. Did the measured mass change after the container was opened? If so, how did it change?

c. In previous lessons, you learned that oxygen gas from the air was used in this reaction. Does oxygen gas have mass? What is your evidence?

d. What do you think happened to cause the measured mass on the scale to change after the container was opened?

e. Draw and label a diagram that shows what you think happened to the amount of matter when the container was opened.

5. Think about your observations of both chemical reactions. Discuss these questions with your team and record your ideas.

   a. After a gas leaves an opened container, is it still being measured by the scale?

   b. When a gas enters an opened container and reacts with the substances in the container, is it part of the initial measurement of the reactants on the scale?

   c. If the measured mass changes during a chemical reaction, what could be the cause? (Remember, the measured mass is the mass of the amount of matter that is measured by the scale.)

   d. Does the *total mass* change during a chemical reaction? (The total mass is the mass of *all* matter involved in the chemical reaction—what is measured and any that *is not* measured by the balance.) Why or why not? Explain your answer.

Pulling It Together

Work on your own to answer these questions. Be prepared for a class discussion.

1. Given your observations, how would you answer the Key Question, **What happens to the amount of matter (mass) when new substances form during chemical reactions?**

2. Think back to the video that shows how nylon forms. Nylon is made from two liquids—hexamethylenediamine and adipic acid. Nylon and water form where the two liquids touch each other.

   a. What do you think will happen to the measured mass of the nylon as more and more of it forms?

   b. What will happen to the measured mass of the liquids in the beaker as more and more nylon thread is pulled out? Why do you think so?

   c. Do you think the measured mass of all the nylon that could be formed would be less than, equal to, or more than the original measured mass of the two liquids? Why do you think so?

3. Given what you learned in previous lessons about chemical reactions, how do you think atoms and molecules could be used to explain the following observations about mass?

   a. Measured mass stays the same when a chemical reaction takes place in a container that is sealed.

   b. Measured mass sometimes changes when a chemical reaction takes place in a container that is open.

Lesson 2.2—Sealed Containers and Total Mass

What do we know and what are we trying to find out?

We saw in Lesson 2.1 that the measured mass does not change when we carry out a chemical reaction in a sealed container even though the products (ending substances) are different from the reactants (starting substances). Because nothing can enter or leave the sealed container, the total amount of the matter of all the substances does not change during chemical reactions that occur in sealed containers.

Remember that all matter is made up of atoms, and we learned in the earlier lessons that atoms are rearranged—not created or destroyed—during chemical reactions. This means that the number of each type of atom stays the same during chemical reactions. In this lesson, you will use models to help you think about what atoms have to do with our observations of mass during chemical reactions that take place in sealed containers.

Think about your experiences representing chemical reactions with models. Then respond to the Key Question to the best of your knowledge.

> **Key Question: Why does rearranging atoms keep the *total* mass constant during chemical reactions?**

Activity 1: Comparing the Mass of LEGO Bricks and Atoms

Materials
For each team of students
LEGO model kit
Model Key

In this activity, you will use the mass of LEGO bricks to help you think about the mass of the atoms they represent.

Procedures and Questions

1. Every type of atom has a unique mass. Scientists have determined the mass of each type of atom. Examine Table 2.3 for the masses of hydrogen, carbon, nitrogen, and oxygen, given in atomic mass units (amu).

Table 2.3. Atom Masses

Atom	Approximate Mass (amu)
Hydrogen	1
Carbon	12
Nitrogen	14
Oxygen	16

2. Discuss the following questions with your team and record your answers.
   a. What do you notice about the mass of a hydrogen atom in comparison with that of a carbon atom?

   b. What do you notice about the mass of a carbon atom in comparison with that of a nitrogen atom or an oxygen atom?

3. With your team, contrast the masses of these LEGO bricks representing different atoms:
   a. A white brick (hydrogen atom) and a black brick (carbon atom)

   b. A black brick (carbon atom), a blue brick (nitrogen atom), and a red brick (oxygen atom)

4. Discuss the following questions with your team and record your answers.
   a. Since there are only two different sizes of LEGO bricks in your kit, how could you use LEGO models to represent four different atoms?

   b. If you were to design a custom LEGO kit, how many different size bricks would you use to represent the four types of atoms in Table 2.3, and what would their relative masses be?

5. Discuss the following questions with your team and record your answers.
   a. In Lessons 1.5 and 1.6, when you built LEGO models of reactant molecules and rearranged the bricks to form product molecules, did the mass of any individual brick change?

   b. Do you think the mass of an actual atom changes during a chemical reaction? Does the mass of an atom change depending on which other atoms it is connected to? Why or why not?

Activity 2: Using Models to Think About Chemical Reactions in Sealed Containers

Materials
For each team of students
LEGO model kit
Chemical Reaction Mats: Steel Wool (Iron) and Air (Oxygen), Baking Soda and Vinegar
Model Key
Access to an electronic scale
Plastic zipper bag

In this activity, you will use models to help you think about what atom rearrangement during chemical reactions has to do with the mass of sealed containers staying constant. Each LEGO brick represents one atom.

Remember, the *Chemical Reaction Mats* are one way to represent the starting and ending substances. They show the following:

Reactants (starting substances) ⟶ Products (ending substances)

The reactants are shown on the green side of the mat. The products are shown on the yellow side of the mat.

Procedures and Questions

1. Use the *Chemical Reaction Mats* and LEGO model kit to complete these steps for the Steel Wool (Iron) and Air (Oxygen) reaction. Fill in Table 2.4 as you work.
   a. In the first row of Table 2.4, write a word equation that represents the Steel Wool (Iron) and Air (Oxygen) reaction.
   b. Use LEGO bricks to build the models of reactants as shown on the green side of the mat. Put all other LEGO bricks aside.
   c. Put the models in the plastic bag, seal it, measure the mass of the models of the reactants with a balance, and record the mass in Table 2.4.
   d. Use only bricks inside the bag to build the models of products as shown on the yellow side of the mat.
   e. Seal the bag again, measure the mass of the models of the products with a balance, and record the mass in Table 2.4.

Table 2.4. Steel Wool (Iron) and Air (Oxygen) Reaction

Reactants ⟶ Products	
Equation:	
Mass (g) of the models of the reactants g	Mass (g) of the models of the products g

2. Repeat Step 1a–e for the Baking Soda and Vinegar reaction. Fill in Table 2.5 as you work.

Table 2.5. Baking Soda and Vinegar Reaction

Reactants ⟶ Products	
Equation:	
Mass (g) of the models of the reactants g	Mass (g) of the models of the products g

3. Think about your experience with the models of both chemical reactions. Discuss the following questions with your team and write down your ideas.
   a. For the Steel Wool (Iron) and Air (Oxygen) reaction, what do you notice about the mass of the products in comparison with the mass of the reactants?
   b. Look back at Lesson 2.1 and see what you wrote in the first two rows of Table 2.2. Use your answer to question 3a above and your observations of the models to explain the data you entered in Table 2.2.
   c. For the Baking Soda and Vinegar reaction, what do you notice about the mass of the products in comparison with the mass of the reactants?
   d. Look back at Lesson 2.1 and see what you wrote in the first two rows of Table 2.1. Use your answer to question 3c above and your observations of the models to explain the data you entered in Table 2.1.
   e. Write a "rule" or general statement that summarizes your findings recorded in 3a–d. Be sure to use ideas about mass and atoms.

Science Ideas

The activities in this lesson were intended to help you construct important ideas about the relationships between atoms and total mass. Read the ideas below. Remember that because science ideas are consistent with evidence from a wide range of phenomena, you can apply them to observations and data about similar phenomena. Look back through Lesson 2.2. In the space provided after each science idea, give evidence that supports the idea.

Science Idea #7: Atoms are not created or destroyed during chemical reactions, so the total number of each type of atom remains the same. We say that *atoms are conserved.*

Evidence:

Science Idea #8: The mass of a particular atom does not change, so a given number of that type of atom will always have the same *total* mass.

Evidence:

Science Idea #9: Because the mass of a particular atom does not change and because the number of each type of atom stays the same, the *total mass* of the matter stays the same even though atoms are rearranged during chemical reactions. We say that *mass is conserved.*

Evidence:

Pulling It Together

Work on your own to answer these questions. Be prepared for a class discussion.

1. After learning about the relationship between atoms and mass, how would you answer the Key Question, **Why does rearranging atoms keep the *total* mass constant during chemical reactions?** Explain it as you would to a friend who missed class.

2. In Lesson 2.1, you observed that measured mass remained constant after baking soda and vinegar reacted in a sealed container. However, the measured mass decreased when the container was opened.

   When you measured the mass of the LEGO models shown on the *Chemical Reaction Mats*, you measured the mass of all of the models of reactants and all of the models of products. This is like measuring the mass of reactants and products of a chemical reaction in a sealed container. Now, let's use the LEGO models to help us think about what happened to the mass when the container was opened.

a. Look at the *Chemical Reaction Mat* for Baking Soda and Vinegar (also shown in Figure 2.1). Circle the model in Figure 2.1 that represents a gas molecule that could leave the opened container.

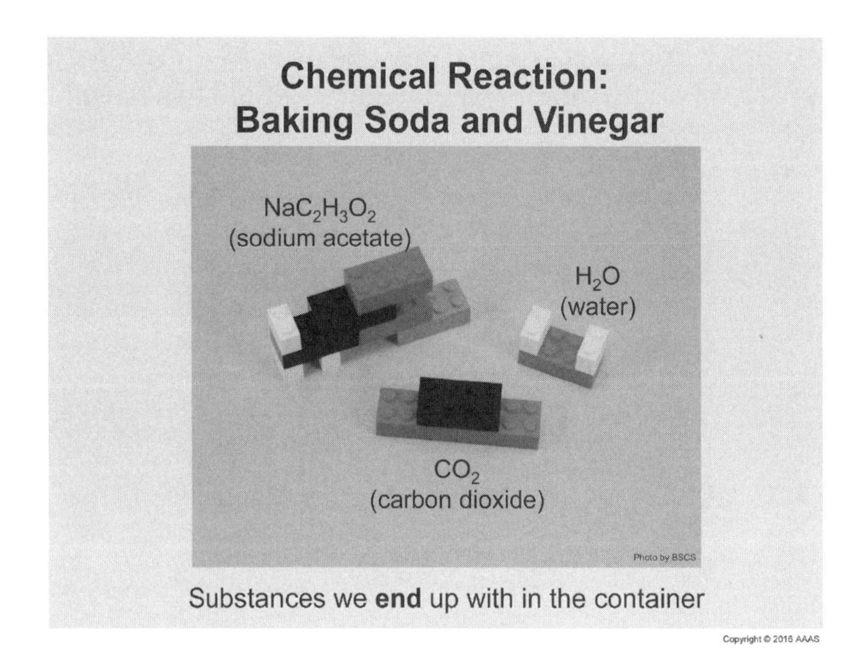

Figure 2.1. *Chemical Reaction Mat*

b. Think about Activity 2, Step 1e, where you measured and recorded the mass of the models of the products with a balance. Imagine that models of all three products (sodium acetate, water, and carbon dioxide) are sitting on a balance.

Describe how you could represent a gas molecule leaving the opened container. Be as specific as you can about which atoms and how many of them will "leave."

c. Table 2.6 shows how many bricks are measured by the balance and the mass of these bricks before the container is opened. Complete the table to show how many bricks are measured by the balance and the mass of these bricks after the gas molecule leaves the opened container.

Table 2.6. Measured Mass of Molecular Models

Bricks, Mass (g)	**Sealed Container**		**Opened Container, After Gas Leaves**	
	# of Bricks on the Balance	**Mass of Bricks on the Balance (g)**	**# of Bricks on the Balance**	**Mass of Bricks on the Balance (g)**
Hydrogen brick, 0.5 g	5	5 bricks x 0.5 g/brick = 2.5 g		___ bricks x 0.5 g/brick = ___ g
Carbon brick, 2 g	3	3 bricks x 2 g/brick = 6 g		___ bricks x 2 g/brick = ___ g
Oxygen brick, 2 g	5	5 bricks x 2 g/brick = 10 g		___ bricks x 2 g/brick = ___ g
Sodium brick, 2 g	1	1 brick x 2 g/brick = 2 g		___ bricks x 2 g/brick = ___ g
Total	14	2.5 g + 6 g + 10 g + 2 g = 20.5 g		___ g + ___ g + ___ g + ___ g = _____ g

d. Did you see a change in the mass of the bricks in the opened container? Which bricks are now gone?

e. If so, does the amount of the change in mass correspond to the mass of the model you circled in the picture in Step 2a? Show your work.

Lesson 2.3—Opened Containers and Measured Mass

What do we know and what are we trying to find out?

The *Chemical Reaction Mats* and other models you used in previous lessons helped you think about what happens when all the reactant molecules are used to form product molecules without any leftover reactants. In the real world, however, there are often more molecules of one reactant than another. When this happens, the chemical reaction will continue until molecules of one of the reactants are used up. Then the reaction will stop.

The *Chemical Reaction Mats* also represent chemical reactions as though all reactants and products are trapped in sealed containers. But you saw in Lessons 2.1 and 2.2 that the measured mass *can* change if the container is opened so that substances can enter or leave.

In this lesson, we will use models to help us think about why measured mass changed in the iron-rusting and baking-soda-plus-vinegar reactions after the containers were opened.

At the end of Lesson 2.2, you used LEGO models to think about how rearranging atoms during chemical reactions conserves atoms and total mass. Use your observations from the models to respond to the Key Question to the best of your knowledge. Write about what would cause measured mass to increase and what would cause measured mass to decrease.

Key Question: If atoms and total mass are always conserved during chemical reactions, why can measured mass change when the container is opened?

Activity 1: Reactants, Products, and Leftovers

Figure 2.2 shows what happens when 10 H_2 molecules and 10 O_2 molecules (left) react to form water molecules (right). You will use this image to help you think about what happens when many reactants are available to undergo a chemical reaction.

Using chemical formulas, the equation for making water from hydrogen gas and oxygen gas is as follows:

$$2\,H_2 + 1\,O_2 \longrightarrow 2\,H_2O$$

In this equation, hydrogen molecules (H_2) and the oxygen molecule (O_2) are shown on the left side of the arrow because they are reactants (starting substances). Water molecules (H_2O) are shown to the right of the arrow because they are the products (ending substance). The equation shows the smallest whole number ratio of reactants and products. That is, when two hydrogen molecules react with one oxygen molecule, two water molecules are made. If four hydrogen molecules react with two oxygen molecules, four water molecules are made, and so forth.

Procedures and Questions

1. Look carefully at Figure 2.2, or at a color version your teacher projects and discuss with your team what you see in the image.

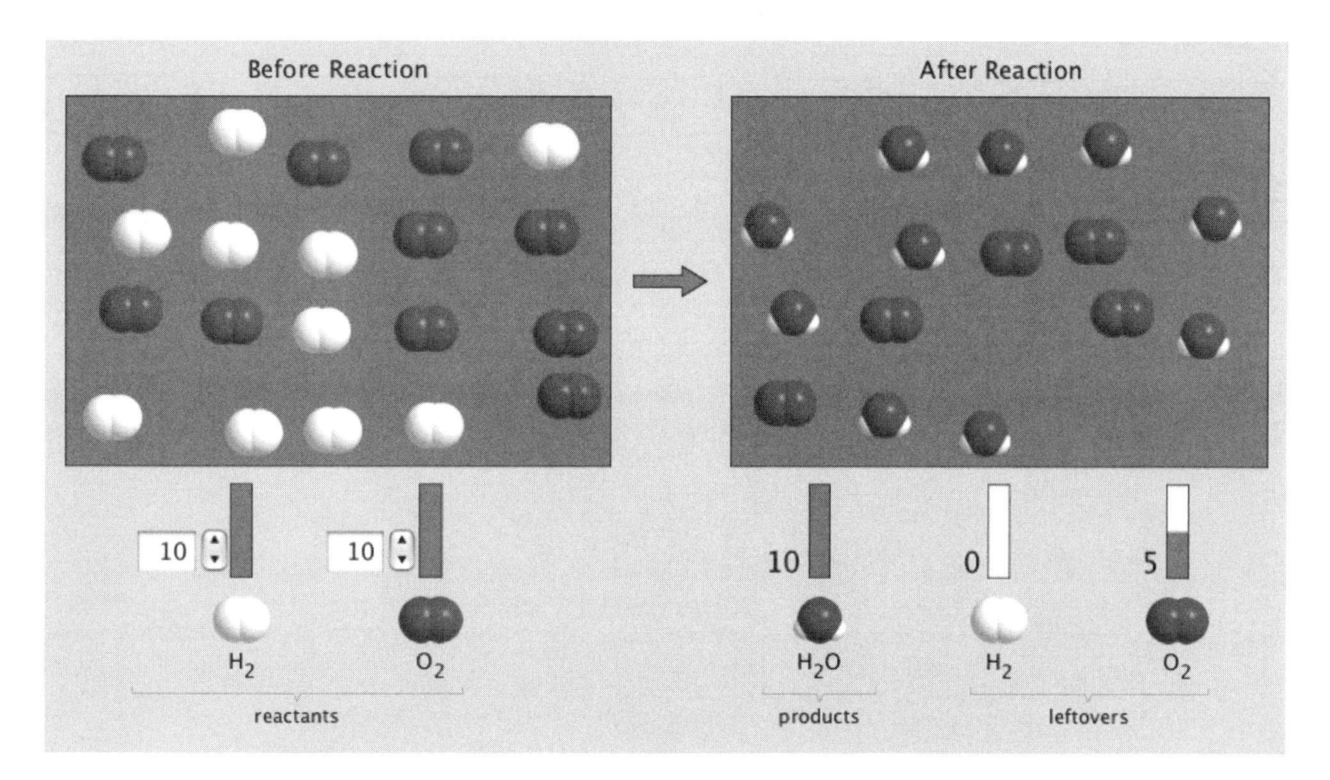

Figure 2.2. Reactants and Products, Before and After Reaction

2. Use Figure 2.2 as you discuss the following questions with your team and record your ideas.

   a. Why can only 10 H_2O molecules be made if there are still 5 O_2 molecules left?

Table 2.7 below shows the number of atoms and molecules represented in Figure 2.2 before and after hydrogen molecules and oxygen molecules react to form water molecules. The number of atoms and molecules has been calculated for some of the cells in the table.

Table 2.7. Comparing Molecules and Atoms of Reactants and Products + Leftover Reactants

		Number of Molecules		Number of Atoms
Reactants	Hydrogen	10	Hydrogen	20
	Oxygen		Oxygen	
	Water			
	Total number of molecules		Total number of atoms	
Products + Leftover Reactants	Hydrogen	0	Hydrogen	20
	Oxygen		Oxygen	
	Water			
	Total number of molecules		Total number of atoms	

3. Complete Table 2.7 and use it as you discuss these questions with your team and record your ideas.

   a. How many *molecules in total* are there before the reaction? How does that compare with the number of molecules there are after the reaction?

   b. After the chemical reaction takes place, how many hydrogen atoms are present? How many oxygen atoms are there?

   c. How does the number of atoms in the reactants of the chemical reaction compare with the number of atoms in the products?

Now that we've looked at the atoms and molecules, let's investigate the mass. Calculate the mass of the reactants and products in Table 2.8. An example has been done for you.

Table 2.8. Comparing Mass of Reactants and Products + Leftover Reactants

Atoms Making Up Reactants	# of Atoms	Mass of 1 Atom	Mass of Atoms in Reaction
Hydrogen	20	1 amu	20 amu
Oxygen			
		Total mass of reactants	
Atoms Making Up Products + Leftover Reactants	**# of Atoms**	**Mass of 1 Atom**	**Mass of Atoms in Reaction**
Hydrogen			
Oxygen			
		Total mass of products	

4. Complete Table 2.8 and use it as you discuss these questions with your team and record your ideas.
   a. What do you notice about the total mass of the atoms before and after the reaction?
   b. What atomic mass did you use to determine the mass of all the hydrogen atoms in products + leftover reactants? Which science idea supports your decision?

5. Answer this question on your own: Is total mass conserved during chemical reactions because atoms are conserved, because molecules are conserved, or because both atoms and molecules are conserved? Use information from Figure 2.2, Table 2.7, and Table 2.8 to support your answer.

Activity 2: Using Models to Think About Chemical Reactions in Opened Containers

Materials
For each team of students
1-gallon zipper plastic bag
LEGO model kit
Chemical Reaction Mats: Steel Wool (Iron) and Air (Oxygen), Baking Soda and Vinegar
Access to an electronic scale

In this activity, you will use models to help you think about why measured mass can change when a container is opened after a chemical reaction has taken place.

First, your teacher will demonstrate what happens when baking soda and vinegar are mixed in a sealed bag and then that bag is opened.

Remember, we have used two different ways to represent chemical equations, one using words and another using symbols. For example, the reaction between baking soda and vinegar can be represented by the following equations:

sodium bicarbonate (baking soda) + acetic acid (vinegar) ⟶ sodium acetate + water + carbon dioxide

$$NaHCO_3 + HC_2H_3O_2 \longrightarrow NaC_2H_3O_2 + H_2O + CO_2$$

Procedures and Questions

1. Watch as your teacher represents the chemical reaction between baking soda and vinegar with models. Record your data in Table 2.9.
   a. Some LEGO models of the molecules of the reactants are placed in a sealed container (the plastic bag, which represents the sealed bottle in Lesson 2.1).
      i. Count and record the number of models of molecules of the reactants in the bag.
      ii. Count and record the number of each kind of atom model in the bag.
      iii. Weigh and record the initial mass of the sealed bag containing the models.
   b. The chemical reaction is modeled in the sealed bag.
      i. Count and record the total number of models of molecules (reactants and products) in the bag.
      ii. Count and record the number of each kind of atom model in the bag.
      iii. Weigh and record the final mass of the sealed bag containing the models.

c. The bag is opened, and models of gas molecules (reactants or products) can enter or leave. In this case, some models of carbon dioxide molecules leave the bag.

i. Count and record the number of models of molecules (reactants and products) of the substances in the bag.

ii. Count and record the number of each kind of atom model in the bag.

iii. Weigh and record the final mass of the opened bag containing the models.

Table 2.9. Baking Soda and Vinegar

	Number of Reactant and Product Models in the Bag	Number of Bricks in the Bag	Mass of the Bag (g)
a.	# $HC_2H_3O_2$ models: # $NaHCO_3$ models: # $NaC_2H_3O_2$ models: # H_2O models: # CO_2 models:	# H bricks: # C bricks: # O bricks: # Na bricks:	Initial mass of sealed bag: g
b.	# $HC_2H_3O_2$ models: # $NaHCO_3$ models: # $NaC_2H_3O_2$ models: # H_2O models: # CO_2 models:	# H bricks: # C bricks: # O bricks: # Na bricks:	Final mass of sealed bag: g
c.	# $HC_2H_3O_2$ models: # $NaHCO_3$ models: # $NaC_2H_3O_2$ models: # H_2O models: # CO_2 models:	# H bricks: # C bricks: # O bricks: # Na bricks:	Final mass after bag is opened: g

2. Discuss these questions with your team and record your ideas.

a. Explain why the measured mass was constant while the bag was sealed.

b. Explain why the measured mass decreased after the bag was opened. Where did the mass go?

Now your team will use LEGO bricks to represent the reaction between iron in steel wool and oxygen in air when steel wool is sealed in a flask.

The reaction between iron and oxygen gas can be represented by the following two models (equations):

$$\text{iron} + \text{oxygen} \longrightarrow \text{iron oxide (rust)}$$

$$4\,Fe + 3\,O_2 \longrightarrow 2\,Fe_2O_3$$

3. Build a model to represent the chemical reaction between iron and oxygen. Record your data in Table 2.10. Build models of seven O_2 molecules. Some of these will represent oxygen molecules in the air inside the sealed container (the bag, which represents the sealed flask). Some will represent the oxygen molecules in the air outside of the container.

   a. Seal some models of the reactant molecules in the container. Use eight Fe atoms and three O_2 molecules. Set aside the other O_2 molecules to represent oxygen molecules in the air outside of the container.

      i. Count and record the number of models of the reactant molecules in the bag.

      ii. Count and record the number of each kind of atom model in the bag.

      iii. Weigh and record the initial mass of the sealed bag containing the models.

   b. Model the reaction in the sealed bag—do not let any matter enter or leave the bag. Make sure that the models of reactants that react are completely used and the models of products are completely formed. That means there should be no parts of molecules.

      i. Count and record the number of models of molecules of the substances.

      ii. Count and record the number of each kind of atom model in the bag.

      iii. Weigh and record the final mass of the sealed bag containing the models.

   c. Open the bag and let the models of gas molecules (reactants or products) enter or leave. In this case, some models of oxygen molecules enter the bag.

      i. Add more O_2 molecules and react them with the Fe atoms. Keep adding O_2 until there are no unreacted Fe atoms left.

      ii. Count and record the number of models of molecules of the substances in the bag.

      iii. Count and record the number of each kind of atom model in the bag.

      iv. Weigh and record the final mass of the opened bag containing the models.

Table 2.10. Steel Wool (Iron) and Air (Oxygen)

	Number of Reactant and Product Models in the Bag	Number of Bricks in the Bag	Mass of the Bag (g)
a.	# Fe bricks: # O_2 models: # Fe_2O_3 models:	# Fe bricks: # O bricks:	Initial mass of sealed bag: g
b.	# Fe bricks: # O_2 models: # Fe_2O_3 models:	# Fe bricks: # O bricks:	Final mass of sealed bag: g
c.	# Fe bricks: # O_2 models: # Fe_2O_3 models:	# Fe bricks: # O bricks:	Final mass after bag is opened: g

4. Discuss these questions with your team and record your ideas.
   a. Explain why the measured mass was constant while the bag was sealed.

   b. Explain why the measured mass increased after the bag was opened. Where did the additional mass come from?

5. Answer these questions individually.
   a. Based on what you observed when you modeled the chemical reaction between baking soda and vinegar and the chemical reaction between iron and oxygen, write a general statement that explains why the measured mass can change when a chemical reaction is carried out in an open system.

   b. Explain why a change in measured mass does not violate ideas about why mass is conserved.

Science Ideas

The activities in this lesson are intended to help you construct an important idea about why measured mass can change. Read the idea below. Notice that Science Idea #10 explains observations about changes in mass in terms of atoms. We can observe when the mass changes, but we can't see atoms entering or leaving the system. However, because Science Idea #10 states a general principle that is consistent with a wide range of observations and data, we can use it to reason that since atoms are not created or destroyed in chemical reactions and the mass of an atom does not change, whenever we see the measured mass change we know that atoms have entered or left the system. You will be expected to use ideas about atoms to explain phenomena involving changes in measured mass.

Look back through Lesson 2.3 so far. In the space provided after the science idea, give at least one piece of evidence that supports the idea.

Science Idea #10: The *measured* mass of reactants and products is not always the same as the *total* mass. The measured mass changes if reactants or products (often gases) enter or leave the system. This is because atoms that make up reactants or products enter or leave the system.

Evidence:

Activity 3: Explaining Changes in Mass

We can use the ideas we have learned in Lessons 2.1, 2.2, and 2.3 to explain how measured mass can change (or not change) during a chemical reaction.

Procedures and Questions

Some students worked together to answer the question below and explain their answer:

> Why did the measured mass increase when steel wool rusted if the total mass of reactants and products stayed the same (was conserved)?

1. Read the explanation, highlight the evidence, and underline the science ideas cited.

Explanation:

The measured mass increased because O atoms from oxygen gas (O_2) entered the container and reacted with Fe atoms from the steel wool to form rust (Fe_2O_3). The measured mass can increase during a chemical reaction if atoms enter the system and react to form molecules that can't leave (Science Idea #10). Atoms have mass, so the more atoms there are in the container, the more its measured mass will be.

Our models showed us that if more O atoms enter the container and react with Fe atoms, more Fe_2O_3 molecules can form and the measured mass can increase. So I think that's what happened with the real atoms. Once the O atoms that entered as oxygen gas were part of the rust, they were no longer a gas and therefore couldn't leave the container.

Evaluate the explanation according to the Explanation Quality Criteria.

2. Use the explanation of why the measured mass increased when iron rusted to help you explain why the measured mass decreased when baking soda reacted with vinegar in an open container. Check to be sure your explanation meets the Explanation Quality Criteria.

Explanation:

3. Why doesn't an increase or a decrease in measured mass contradict the science idea that matter is conserved?

Pulling It Together

Work on your own to answer these questions. Be prepared for a class discussion.

1. Now that you have modeled chemical reactions in opened containers, how would you answer the Key Question, **If atoms and total mass are always conserved during chemical reactions, why can measured mass change when the container is opened?** Explain it as you would to a friend who missed class today.

   Explain both cases—why measured mass sometimes increases and why it sometimes decreases when the container is opened.

2. Think back to the video you saw in Chapter 1 of how nylon is made.

   a. If all of the liquid reactants in the beaker were used to make nylon thread and water, would the number of atoms in the nylon and water be less than, the same as, or more than the number of atoms originally in the beaker? Explain why.

   b. As the nylon thread is pulled out of the liquids in the beaker, the measured mass of the liquids in the beaker decreases. Explain why in terms of atoms.

3. When a log burns in a fireplace, two gases (carbon dioxide and water vapor) are made. Imagine that you and a friend are talking about what happens when a log is burned in a fireplace. She thinks that the matter that makes up the log is destroyed. She thinks this because the log weighs much more than the ashes that are left over after the log burns.

   a. You and your class complete the following table to help decide whether your friend is right. In the second column, fill in science ideas, evidence, or models that support the statements in the first column.

   For each supporting item that you write, indicate whether it states the science idea, provides evidence, uses the science idea, or can be shown with a modeling activity. Hint: Some of the boxes will have more than one type of support for the statement.

Carbon dioxide and water vapor have different properties from the log.	
Because carbon dioxide and water have different properties from the log, a chemical reaction must have occurred.	
Carbon dioxide and water vapor have different properties from the log because they are made of different arrangements of atoms.	
In chemical reactions, atoms of the starting substances disconnect and reconnect in new ways to form ending substances.	
Atoms are not created or destroyed in chemical reactions.	
Because the atoms just rearranged, the total mass after the reaction is the same as the total mass before the reaction.	

Statement that will help explain your answer	Science idea, evidence, or modeling activity that supports the statement
Carbon dioxide and water vapor (the ending substances) are gases that can leave open systems.	
The fireplace is an open system, so gases can leave the system.	

b. Based on the table, what would you tell your friend?

4. The Statue of Liberty is made up of copper (Cu atoms). But the statue doesn't have the shiny, orange-brown appearance of copper. Instead, it is green. Why? After being exposed for many years to oxygen and carbon dioxide in the air, a thin layer of green copper carbonate ($CuCO_3$ molecules) formed on the copper statue. Answer the following questions, being sure to use ideas about atoms in your explanation and to write an explanation that meets the Explanation Quality Criteria.

   a. Is the change a chemical reaction? Explain.

   b. Do you think the Statue of Liberty now has less mass than, the same mass as, or more mass than when it was first made? Explain.

Toward High School Biology

Chapter 3

Student Edition

Lesson 3.1—The "Stuff" That Makes Up Plants

What do we know and what are we trying to find out?

In the previous two chapters, you observed and modeled chemical reactions in nonliving systems because they were simpler systems than the bodies of plants and animals. By observing and modeling chemical reactions in simpler systems, you were able to figure out what happens to atoms and molecules during chemical reactions and to the measured mass of the system. You are now ready to apply what you learned about chemical reactions to the growth of plants. In order to think about how plants might grow, you need to know what plant body structures are made of. That will give you clues to what substances plants will need in order to grow.

Answer the Key Question to the best of your knowledge. Be prepared to share your ideas with the class.

> **Key Question: What substances do plants need in order to grow and repair?**

Activity 1: Observing Plant Growth

To help us think about how plants grow, we'll start by observing examples of plant growth. As you watch the video and slides, think about the body structures being made as the plant grows or repairs itself after an injury.

Procedures and Questions

1. Observe the video and photos of plant growth. As you answer the questions below, think about the body structures plants have that you can see and also the body structures you can't see.

   a. What body structures are getting bigger as the **corn plants** grow?

   b. What body structures are getting bigger as the **pumpkin** grows?

   c. What body structures are getting bigger as the **kudzu vines** grow?

   d. What body structures are getting bigger as the **giant sequoia trees** grow?

   e. What body structures are being rebuilt as the **tree heals a wound** after one of its branches was pruned (chopped off)?

2. You have just observed that various plant body structures get bigger as plants grow, but how do we know that plants are actually adding new matter to their bodies? To answer this question, we will observe what happens to the mass of growing plants. Examine the data in Table 3.1 and answer the questions that follow.

Table 3.1. Typical Change in Mass (by Weight in Grams) of Some Growing Plants, Over Different Time Periods

Plant	Initial Weight (Age)	Final Weight (Age)	Change in Weight	Increase or Decrease?
Mung plants	24.6 g (1 week)	28.1 g (3 weeks)		
Bean plants	47 g (3 weeks)	262 g (10 weeks)		
Corn plants	0.7 g (1 week)	896 g (13 weeks)		
Grass	2 g (60 days)	13 g (140 days)		

a. Using the data in Table 3.1, calculate the change in weight for each plant (you can use a calculator). Put a plus sign (+) next to any change in mass that was an increase. Put a minus sign (–) next to any change in mass that was a decrease.

b. Based on these data, what happens over time to the mass of a plant when it grows?

Activity 2: The Substances That Make Up Plants

The video and slides you observed in Activity 1 showed that plants produce a lot of different body structures as they grow. What substances are these body structures made up of? When you eat fruit like strawberries or tomatoes, or stems like celery or bamboo shoots, you have probably observed that lots of liquid comes out of them. Plant body structures are made up of a lot of water. But water doesn't give tomatoes or celery their texture, flavor, or nutritional value. If you have ever eaten packaged vegetables, you may have noticed the label on the package. The label indicates the amounts of the main nutrients—carbohydrates, fats, and proteins—plus a few other substances like sodium, cholesterol, and vitamins in a serving of the food. In this activity, you will find out how much of each of the main nutrients make up some common plant body structures.

Nutrition Facts

Serving Size 1 cup raw green beans (100g)
Servings Per Container 1

Amount Per Serving	
Calories 30	Calories from Fat 0
	% Daily Value*
Total Fat 0g	0%
Saturated Fat 0g	0%
Trans Fat 0g	
Cholesterol 0mg	0%
Sodium 5mg	0%
Total Carbohydrate 7g	2%
Dietary Fiber 3g	12%
Sugars 3g	
Protein 2g	

Procedures and Questions

1. Observe the slides your teacher will show you of plant body structures.

2. Examine Table 3.2 with your group and answer the questions that follow.

Table 3.2. Relative Mass (in Grams per 100 Grams) of Carbohydrate Molecules, Fat Molecules, and Protein Molecules in Some Plant Body Structures

Structure	Plant	Carbohydrates (per 100 g)	Fats (per 100 g)	Proteins (per 100 g)
Root	Carrot	8.24	0.13	0.64
	Potato	17.00	0.10	2.00
Stem	Bamboo shoot	5.20	0.30	2.60
	Celery	2.97	0.17	0.69
Leaf	Spinach	3.63	0.39	2.86
	Lettuce	2.87	0.15	1.36
Flower	Broccoli	5.24	0.35	2.98
	Cauliflower	4.97	0.28	1.92
Fruit	Strawberry	7.68	0.30	0.67
	Tomato	3.89	0.20	0.88
Seed	Snap beans	6.97	0.22	1.83
	Kidney beans	4.10	0.50	4.20

a. In Table 3.2, **circle** the type of molecule that is present in the largest amount in most of the plants' body structures.

b. Are there any plant structures for which the molecule you circled is not present in the largest amount? If so, which one(s)?

c. Based on the data in Table 3.2, what type of molecule do you think plants will need the most of in order to make their bodies? Why do you think so?

Pulling It Together

Work on your own to answer these questions. Be prepared for a class discussion.

1. Now that you have seen plants grow and learned about the substances that make up their body structures, how would you answer the Key Question, **What substances do plants need in order to grow and repair?**

2. Where do you think the atoms for plant growth come from? List all the places you think plants might get atoms from.

Lesson 3.2—Carbohydrates That Make Up Plants

What do we know and what are we trying to find out?

We saw in the last lesson that the different body structures of plants are mostly made up of carbohydrates, or "carbs," as they're often called. So to understand how plants grow, we'll need to take a closer look at the carbohydrates that make up plants. Think about the last time you bit into a ripe fruit, such as a tomato or strawberry. Did it taste sweet? Now compare the taste of the ripe tomato or strawberry with the taste of a stem of celery or a root vegetable, such as a potato. Were they as sweet? Next, compare what it feels like when you bite into a strawberry with what it feels like when you bite into a piece of celery. What do you think it would feel like if you tried to bite into the bark of a tree? All of these different plant structures are mostly carbohydrates, yet they are not all the same. In this lesson, we will examine ways that carbohydrates are similar to one another and how they differ.

Answer the Key Question to the best of your current knowledge. Be prepared to share your ideas with the class.

Key Question: Are the carbohydrates that make up different plant structures the same substance?

Activity 1: Properties of Familiar Carbohydrates

Materials
For each team of students
1 celery stalk

You learned in Chapter 1 that the way to tell whether two substances were the same or different was to look at their properties. In this activity, you will examine the properties of three carbohydrates to decide whether any two are the same substance and consider what plants might use them for.

Procedures and Questions

1. Examine Table 3.3, and then work with your group to answer the questions that follow.

Table 3.3. Properties of Three Common Carbohydrate Molecules

Carbohydrate Molecule	Solubility in Water	Does It Form Fibers?
Glucose	High	No
Starch	Low	No
Cellulose	Low	Yes

   a. Based on the data in Table 3.3, is glucose a different substance from cellulose? Why or why not?

   b. Is cellulose a different substance from starch? Why or why not?

2. Your teacher will hand out a celery stalk. Examine it with your group. Try pulling it apart in different directions or peeling off sections of it. The strands that come off are fibers.

   a. Which of the carbohydrate molecules in Table 3.3 do you think gives celery its stringiness? Why do you think so?

b. Do you think that molecules that form fibers or those that do not form fibers would be better for building some plant structures? Why do you think so?

3. Observe your teacher demonstrate adding pure glucose to water.

   a. Record your observations of what happens below. Do your observations match the data in the table?

   b. Would glucose be good for building plant body structures? Why or why not?

Activity 2: Molecular Composition and Structure of Carbohydrates

You learned in Chapter 1 that if two substances have different properties, then they are made up of different molecules. And you learned that molecules can differ in the types of atoms that make them up. In this activity, you will have a chance to examine models of the molecules making up the three carbohydrates you just decided were different substances—glucose, starch, and cellulose.

Procedures and Questions

1. Examine the data in Table 3.4. What does it tell you about the types of atoms making up the three carbohydrates?

Table 3.4. Composition of Common Carbohydrates

Carbohydrate	% C Atoms	% H Atoms	% O Atoms	% Other Atoms	Total
Glucose	25.0	50.0	25.0	0	100
Starch	28.6	47.6	23.8	0	100
Cellulose	28.6	47.6	23.8	0	100

   a. If glucose, starch, and cellulose are all made up of the same types of atoms, how might they differ?

2. In Figures 3.1, 3.2 (p. 112), and 3.3 (p. 112), color the O atoms red. With your group members, carefully examine the figures and answer the questions that follow.

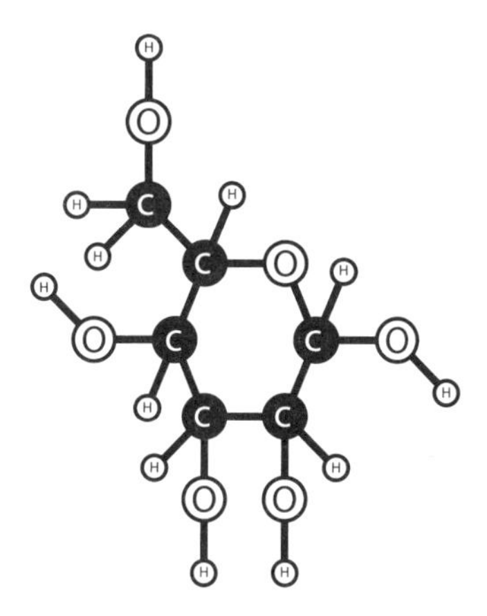

Figure 3.1. Model of the Structure of a Carbohydrate Molecule Called ***Glucose***

Some polymers, like starch, can also have branches coming off the main chain.

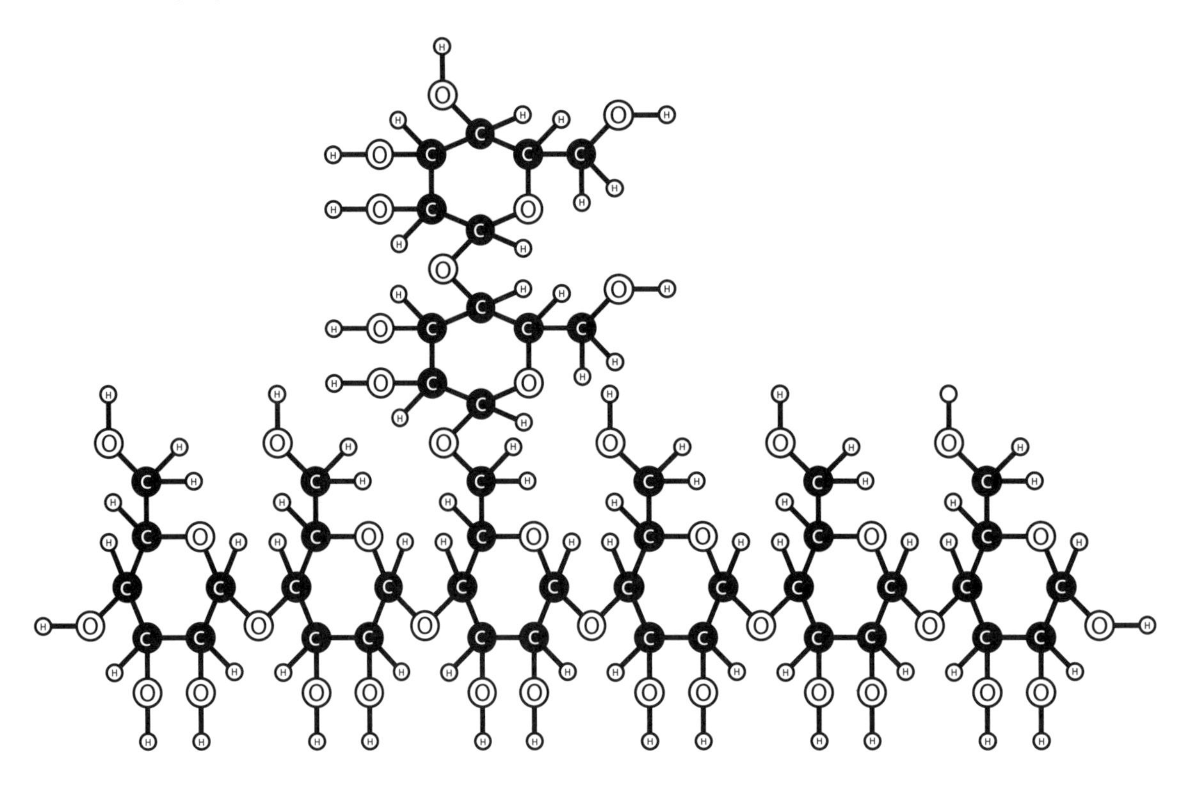

Figure 3.2. Model of the Structure of a Small Piece of a Carbohydrate Molecule Called ***Starch***

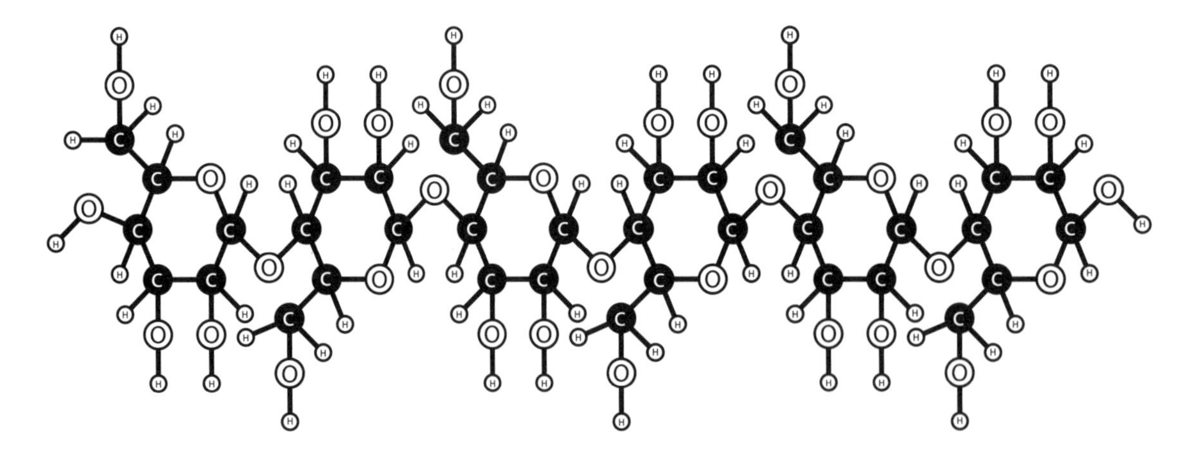

Figure 3.3. Model of the Structure of a Small Piece of a Carbohydrate Molecule Called ***Cellulose***

a. Look at the glucose molecule in Figure 3.1. Count the number of carbon, hydrogen, and oxygen atoms it is made up of, and use your counts to write the chemical formula for glucose. Hint: The atoms are in the order CHO in glucose.

b. Now look at the small piece of the starch molecule in Figure 3.2 and identify the repeating unit. Circle one repeating unit and write down how many units are shown in the piece of starch molecule. What did we call the repeating units of nylon? What did we learn to call a molecule made up of repeating units?

c. Now look at the piece of the cellulose molecule in Figure 3.3. Is cellulose also a polymer? If so, circle one repeating unit and write down how many are shown. Are they the same or different in their orientation (position)?

d. Do you think the glucose molecule in Figure 3.1 could be the monomer used to build the carbohydrate polymers starch and cellulose? Why or why not?

Science Ideas

Lessons 3.1 and 3.2 were intended to help you understand some important ideas about carbohydrate molecules. Read the idea below. Remember that because science ideas are consistent with evidence from a wide range of phenomena, you are justified in using them to explain similar phenomena. Look back through Lessons 3.1 and 3.2 and describe evidence that illustrates each part of the science idea.

Science Idea #11: Carbohydrates are the main polymers making up plant body structures. Carbohydrate polymers are molecules made of glucose monomers. Different carbohydrate polymers have different properties because they are made of different numbers and arrangements of glucose monomers.

Evidence:

Pulling It Together

Answer individually and be prepared for a class discussion.

1. Now that you know more about carbohydrates, how would you answer the Key Question, **Are the carbohydrates that make up different plant structures the same substance?** Provide as much evidence as you can.

2. What chemical reaction do you think might happen when plants build cellulose polymers? Write the word equation for your proposed reaction below, and give a reason for why you think so.

Lesson 3.3—Making Glucose in Plants

What do we know and what are we trying to find out?

We now know that plant body structures are made up mostly of carbohydrate polymers like cellulose and starch. The molecular structures of these polymers suggest that they could be made from glucose monomers. But where do the glucose monomers come from? Do plants "eat" glucose monomers or do they make them? It turns out that there is almost no glucose in the soil or water. Bacteria in the soil and water devour any glucose that finds its way there. So if plants need glucose monomers, they probably have to make them. This lesson will help you figure out how.

Answer the Key Question to the best of your current knowledge. Be prepared to share your ideas with the class.

Key Question: How can plants get the glucose monomers they need for building carbohydrate polymers?

Activity 1: Substances That Plants Take In From Their Environment

In the previous lesson, we saw that glucose is made up of carbon, hydrogen, and oxygen atoms. But where do these atoms come from? To figure that out, we might start by looking at what substances plants take in from their environment that could serve as a source of these atoms.

Procedures and Questions

1. Examine the information in Table 3.5 and answer the questions that follow.

Table 3.5. Substances Plants Take In From the Environment

Substance	Atomic Composition	Location in Environment	Plant Structure That Takes It in
Water (H_2O)	H, O	Ground	Roots
Carbon dioxide (CO_2)	C, O	Air	Tiny holes in the leaves
Minerals (in various molecular forms)	Mainly N, K, Ca, P, Mg, S*	Ground	Roots

**N = Sodium, K = Potassium, Ca = Calcium, P = Phosphorus, Mg = Magnesium, S = Sulfur*

a. Based on what you know about the molecular composition of glucose, which of the substances in the table could supply the needed atoms?

b. Will any of the minerals plants take in from the environment become part of glucose? Why or why not?

Activity 2: Can Plants Make Glucose From Carbon Dioxide and Water?

In the last activity, you decided that carbon dioxide and water molecules could supply plants with the atoms needed to make glucose. In this activity, you will use data from two experiments that scientists carried out to determine whether plants actually do make glucose from carbon dioxide and water. The experiments were first carried out in the 1940s using green algae. Later experiments gave the same results when carried out with plants that live on land or in the water.

Procedures and Questions

1. The experiments used a technique called *isotopic labeling* to track atoms from starting substances to ending substances.

   Isotopic labeling takes advantage of the fact that some atoms of the same type happen to be a little different from the regular atoms. These different "versions" of the atoms (called *isotopes*) have different masses but behave the same in chemical reactions. Some isotopes are unstable (radioactive) and break apart into fragments that can be detected with special instruments. For example, an unstable isotope of carbon has a mass of 14 amu and is referred to as carbon-14 or ^{14}C. Other isotopes are stable and must be detected by their masses. For example, a stable isotope of an oxygen atom has a mass of 18 amu. What do you think it is called?

   You will learn more about isotopes of atoms in high school chemistry, but two things are important for now. First, isotopes of atoms undergo the same chemical reactions as the regular versions, so they rearrange to form the same new molecules just like the regular versions do. Second, scientists can prepare molecules of starting substances that have some of the isotopic atoms. This allows scientists to use their instruments to see which molecules have the isotopes at the end of the chemical reaction. In this way, scientists can figure out what the starting and ending substances are for chemical reactions that are ocurring in the bodies of living organisms.

   Let's look at two experiments that scientists carried out to try to find out where the carbon, hydrogen, and oxygen atoms of glucose come from. For each experiment, look carefully at the data in the table, and then work with your group to answer the questions.

2. Experiment 1: In the first experiment, scientists used carbon dioxide made with labeled carbon atoms so that they would be able to see where the carbon atoms ended up after the carbon dioxide and water molecules reacted. Table 3.6 summarizes their data.

Table 3.6. Location of Labeled Atoms (highlighted) at the Beginning and End of Experiment 1

	Starting Substances (Reactants)		Ending Substances (Products)	
Experiment 1	**C**O_2	H_2O	**C**$_6H_{12}O_6$	O_2

a. What can scientists conclude from Experiment 1 about the carbon atoms? What piece of data can they cite as evidence for their conclusion?

Scientists can conclude that the C atoms in glucose came from carbon dioxide. The evidence they can cite is the observation that the labeled C atoms that started out in carbon dioxide ended up in glucose.

b. Does the data provide evidence for a conclusion about where the oxygen atoms from carbon dioxide end up? Why or why not?

No, because they didn't track the O atoms. The O atoms could have come from carbon dioxide or water. We don't know yet.

3. Experiment 2: In the second experiment, scientists used water made with labeled oxygen atoms and looked to see where the oxygen atoms ended up after water and carbon dioxide reacted. Table 3.7 summarizes their data.

Table 3.7. Location of Labeled Atoms (highlighted) at the Beginning and End of Experiment 2

	Starting Substances (Reactants)		Ending Substances (Products)	
Experiment 2	CO_2	H_2**O**	$C_6H_{12}O_6$	**O**$_2$

a. What can scientists conclude from Experiment 2? What piece of data can they cite as evidence for their conclusion?

Scientists can conclude that the O atoms in water are the same O atoms as are found in oxygen gas. They can cite Experiment 2 as evidence, and how the isotopically labeled oxygen was later found in oxygen gas.

b. What can scientists conclude about where the O atoms that make up glucose come from?

c. What evidence can they cite for this conclusion?

4. Write the word equation for the chemical reaction that plants use to make glucose.

5. Write the chemical formulas for each starting and ending substance:

Starting Substances	Ending Substances

6. Draw a diagram to illustrate the reaction in Experiments 1 and 2. Based on data from the experiments, use arrows to show where the C and O atoms are located in the reactants and where they end up in the products of the reaction.

Activity 3: Modeling the Chemical Reaction That Plants Use to Make Glucose

Materials
For each team of students
Ball-and-stick model kit
Chemical Reaction Mat: Making Glucose
Tray for working with models on desks
Model Key

In the last activity, you determined that carbon dioxide and water are the starting substances for the chemical reaction that produces glucose and oxygen, and that carbon dioxide provides the C atoms for making glucose and probably the O atoms too. In this activity, you will figure out how many carbon dioxide molecules and how many water molecules are needed to make each glucose molecule. Then you will model the chemical reaction to see if glucose and oxygen molecules can be made by rearranging atoms from those carbon dioxide and water molecules. You should also be able to figure out how many oxygen molecules will be produced.

Procedures and Questions

1. Use the information in Table 3.6 of Activity 2 to figure out how many carbon dioxide molecules and water molecules are needed to make a molecule of glucose. Record your answers in the table below.

How many carbon dioxide molecules are needed to make a molecule of glucose?	
How many water molecules are needed to make a molecule of glucose?	

2. Examine the green side of your *Chemical Reaction Mat* to check your answers. Use your ball-and-stick model kit to build the reactant molecules. Depending on the size of your team, each member will need to make 3 or 4 molecules.

3. Close the ball-and-stick kit and put it aside.

4. Flip your mat to the yellow side.

5. Work together to build the product molecules on the yellow side of the mat from the reactant molecules.

6. Record the number of molecules your team has made:

Number of glucose molecules	
Number of oxygen molecules	

7. With your team, answer the following questions.

   a. Write the word equation for the reaction you just modeled. This chemical reaction is called *photosynthesis*.

   b. Write the equation for the chemical reaction you modeled, using the chemical formula for each reactant and product. In front of each reactant, write how many molecules of it were involved in the reaction. Then in front of each product, write how many were produced.

   c. Are atoms rearranged during this chemical reaction? Use models to support your answer.

   d. How does the number of carbon, hydrogen, and oxygen atoms in the reactants compare with the number of carbon, hydrogen, and oxygen atoms in the products? Are atoms conserved during this chemical reaction?

8. Carefully take apart your models and put the pieces back in the bag for the next class to use.

Science Ideas

Lesson 3.3 was intended to help you understand an important idea about how plants make glucose. Read the idea below. Remember that because science ideas are consistent with a wide range of relevant evidence, you are justified in applying them to new observations and data. Look back through Lesson 3.3. Give at least one piece of evidence from your work so far and explain how it supports this idea.

Science Idea #12: Plants use carbon dioxide and water molecules in their environment to make glucose and oxygen molecules. Atoms are rearranged and conserved during this chemical reaction.

Evidence:

Pulling It Together

1. Now that you know more about glucose, how would you answer the Key Question, **How can plants get the glucose monomers they need for building carbohydrate polymers?**

2. Recall the reaction that formed rust. How is photosynthesis similar to the formation of rust? Think about a plant's body as being like the container that held the steel wool. Write about what you think happens to both measured mass and total mass of the plant.

3. The owner of a plant store wanted to see if the amount of carbon dioxide in the air would affect how her plants grow. She set up two identical rooms, Room 1 and Room 2, in her greenhouse. She put one plant in each room. Both plants were the same size and mass, and both were given plenty of water. Room 1 had normal air with the normal level of carbon dioxide in it. Room 2 was filled with air that had extra carbon dioxide in it. See Figure 3.4 below, or the color version your teacher projects.

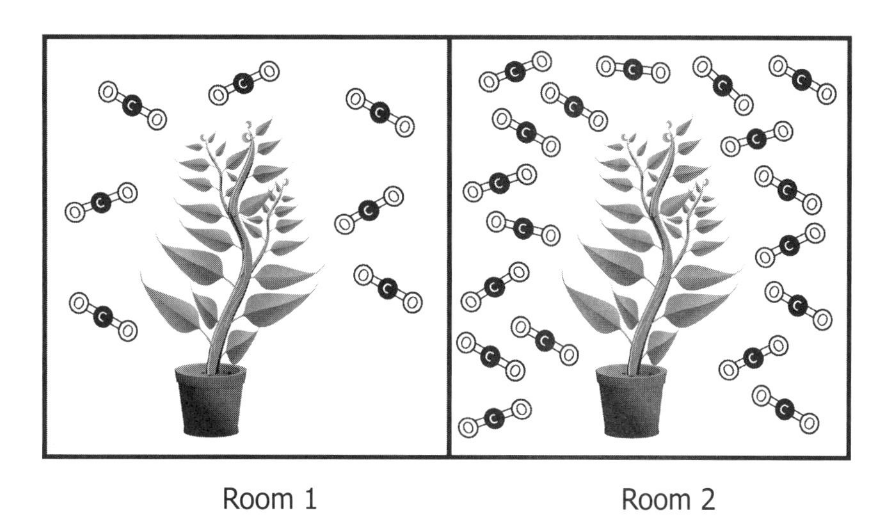

Figure 3.4. Plants Exposed to Different Amounts of Carbon Dioxide

Three months later, the store owner measured each of the plants. Do you think one plant was bigger than the other? Explain your answer using science ideas, evidence, and models. Check to be sure that your explanation meets the Explanation Quality Criteria.

Lesson 3.4—Making Cellulose Polymers in Plants

What do we know and what are we trying to find out?

We know that plant body structures are mostly made up of carbohydrate molecules, and that some carbohydrate molecules, like cellulose, are better for building body structures than others. When we examined the model of a piece of cellulose, we thought it might be made up of glucose monomers. But while models can give us ideas about how something might work, they don't provide evidence for claims. If we want evidence that plants actually make cellulose polymers from glucose monomers, we need to collect data. In this lesson, we will examine experimental data to see if it supports our claim or idea that plants make cellulose polymers from glucose monomers.

Answer the Key Question to the best of your current knowledge. Be prepared to share your ideas with the class.

Key Question: How do plants make the cellulose polymers they need to build their body structures?

Activity 1: Following "Labeled" Carbon in Glucose

In this activity, you will again take advantage of the isotopic labeling method to help you figure out whether growing plants use glucose to make cellulose polymers. In the experiment described below, scientists were trying to stop plants from growing. However, the data they collected also provide evidence for how growing plants use glucose to build their body structures.

Procedures and Questions

1. Watch as your teacher projects information about herbicides.

2. Read about the experiment that scientists conducted, examine the data they collected, and then answer the questions that follow with your group.

Effects of the Herbicide Triazofenamide on Plant Growth

Introduction

Herbicides are substances that stop or prevent the growth of weeds (unwanted plants). In agriculture, weeds take away resources like water and space from desired plants or crops. Herbicides prevent weeds from growing, so farmers often spray their fields with them. There are many different herbicides that are commonly used. Scientists wanted to know if the herbicide triazofenamide also stopped the growth of the surrounding plants.

Experiment

Mouse-ear cress plants were split into two groups: **experimental** and **control**. The experimental group was sprayed with the herbicide, and the control group was not. Although plants normally make their own glucose, scientists gave the plants glucose molecules that had labeled carbon atoms, and then measured how many of the labeled carbon atoms ended up in cellulose in both experimental and control plants. The amount of labeled carbon atoms in cellulose would tell scientists how much cellulose was produced by each group.

Results

Table 1. Amount of labeled carbon atoms in cellulose in dpm (a unit that indicates the level of radioactivity in a substance or material) for each group in the experiment

Group	Amount of Labeled C Atoms in Cellulose (dpm)
Control	11,200
Experimental (herbicide)	1,900

Heim, D. R., I. M. Larrinua, M. G. Murdoch, and J. L. Roberts. 1998. Triazofenamide is a cellulose biosynthesis inhibitor. *Pesticide Biochemistry and Physiology* 59 (3): 163–168.

a. Based on the data in Table 1, can mouse-ear cress plants make cellulose from glucose? What evidence supports your answer/claim?

b. Do the data show that the herbicide prevents mouse-ear cress plants from making cellulose from glucose? Why or why not?

c. Do the data show that the herbicide prevents mouse-ear cress plants from carrying out photosynthesis (making glucose)? Why or why not?

d. How do you predict the mass of the experimental plants would compare with the mass of the control plants after several weeks? Explain your prediction.

Activity 2: Modeling the Reaction That Plants Use to Build Cellulose

Materials
For each team of students
Two premade models of glucose molecules (with stickers on one or more carbon atoms)
Chemical Reaction Mat: Making Cellulose
Tray for working with models on desks
Model Key

So we now know that mouse-ear cress plants can produce cellulose polymers from glucose monomers. Do any atoms rearrange during the process? Let's model the reaction to get ideas about whether any atoms form different connections (rearrange) and, if so, which ones.

Procedures and Questions

1. With your group, look at the molecule models you have been given, and then look at the yellow side of the *Chemical Reaction Mat*. How many new connections will form?
2. On the green side of the mat, draw a line between the atoms that will connect to form the piece of cellulose polymer.
3. Flip your *Chemical Reaction Mat* to the yellow side.
4. Using only the reactant models, build models of the products as shown on the yellow side of the mat. **Do not take apart the models you have just built—leave them for the next class.**
5. With your team, answer the following questions.
   a. Did any atoms rearrange when you modeled this reaction? If so, which atoms rearranged?
   b. Write a word equation for the reaction you just modeled.
   c. When this reaction happens in a plant, what must be happening to the mass of the plant's body structures? Justify your answer.

Science Ideas

Lesson 3.4 was intended to help you understand some important ideas about how plants grow. Read the ideas below. Notice that Science Ideas #13 and #14 explain observations of plant growth in terms of atoms. We can observe when plants grow, but we can't see atoms rearranging or entering or leaving the system. However, because Science Ideas #13 and #14 state general principles that are consistent with a wide range of observations and data, we can use them to reason about the growth of all plants. You will be expected to use ideas about atoms to explain phenomena involving plant growth.

Look back through Lesson 3.4 and give at least one piece of evidence for each idea.

Science Idea #13: Plants use glucose monomers to make cellulose polymers and water molecules. Atoms are rearranged during this chemical reaction.

Evidence:

Science Idea #14: When plants grow or repair, they increase in mass. This increase in measured mass comes from the incorporation of atoms from molecules that were originally outside of the plants' bodies.

Evidence:

Pulling It Together

1. Now that you've learned more about cellulose, how would you answer the Key Question, **How do plants make the cellulose polymers they need to build their body structures?** Start with glucose.

2. A friend says that "living things, like people and trees and dogs, are very different from nonliving things like rocks, water, and air. Nonliving matter could never become part of a living thing." Is he correct or incorrect? What is your evidence?

3. How is plant growth similar to nylon formation?

4. Look back at Lesson 3.2, Figure 3.2 (p. 112). How do you think plants might make starch molecules, starting with glucose molecules?

Lesson 3.5—Explaining Where the Mass of Growing Plants Comes From

What do we know and what are we trying to find out?

In Chapter 3, you have investigated how plants grow. You have applied what you learned in Chapter 1 about atom rearrangement during chemical reactions to the growth of plants. In this lesson, you will apply what you learned in Chapter 2 about atom conservation, conservation of total mass, and changes in measured mass to plant growth. Answer the Key Question to the best of your knowledge. Be prepared to share your ideas with the class.

Key Question: Where does the mass of growing plants come from?

Activity 1: Atoms That Contribute Mass to Plant Body Structures

You learned in earlier lessons that plant body structures are made up mainly of carbohydrate polymers such as cellulose and starch and that cellulose and starch are made up of C, H, and O atoms. So does that mean that the mass of a plant is mainly due to the C, H, and O atoms that make up its body structures? If so, do C, H, and O atoms contribute equal amounts to the mass of plants? In this activity, you will examine data to answer these questions.

Procedures and Questions

1. Examine the information in Table 3.8 and answer the questions that follow.

Table 3.8. Percentage of Dry Weight by Type of Atom for an Evergreen Tree

Atom Name(s)	Symbol(s)	% of Dry Weight
Carbon, hydrogen, oxygen (combined)	C, H, O	96.0
Nitrogen	N	1.3–3.5
Potassium	K	0.7–2.5
Calcium	Ca	0.3–1.0
Phosphorus	P	0.2–0.6
Magnesium	Mg	0.1–0.3
Sulfur	S	0.1–0.2

a. According to Table 3.8, which atoms make up 96% of an evergreen tree's dry mass?

b. Based on the data in Table 3.8, how does the amount of minerals (N, K, Ca, P, Mg, S) compare with the amount of C, H, and O atoms that make up an evergreen tree?

c. How much do you think minerals contribute to the tree's increase in mass as it grows? Why?

d. You have probably heard that plants need sunlight in order to grow. This is true; however, sunlight is not made up of atoms. If sunlight isn't made up of atoms, would it contribute any mass to a growing plant? Why or why not?

2. In Lesson 3.3, you determined that the C and O atoms that make up glucose come from CO_2 and that the H atoms come from H_2O. The chemical formula for glucose, $C_6H_{12}O_6$, tells you that glucose is made up of 6 C atoms, 6 O atoms, and 12 H atoms. Which atoms contribute most of the mass of glucose? If all types of atoms had the same mass, then you could figure it out by counting atoms. However, you already know that different types of atoms have different masses. So to answer the question, you need to find out the mass of each type of atom.

   a. Look back at Table 2.3 (p. 78) to find the mass of H and O atoms, and record the information in Table 3.9 below. The mass of C has already been filled in for you.

   b. Using the first row of Table 3.9 as a model, fill in the missing information for H and O, and then calculate the percentage of the total mass of glucose that comes from H and O atoms.

Table 3.9. Composition of Glucose

Type of Atom in Glucose	Mass of Each Type of Atom (amu)	Number of Each Type of Atom in Glucose	Mass (amu) of Atoms of Each Type in Glucose	% of Total Mass of Glucose
Carbon, C	12 amu	6 atoms	12 amu/atom x 6 atoms = 72 amu	72 amu/180 amu = 40%
Hydrogen, H				
Oxygen, O				
			Total mass = 180 amu	100%

3. Based on Tables 3.8 and 3.9, how much of an evergreen tree's dry weight is due to C and O atoms? Some? Most? Almost all? Why do you think so?

Activity 2: Analyzing, Critiquing, and Revising Explanations

Science is a way of explaining the natural world. Sooner or later, the validity of scientific conclusions is settled by referring to observations of phenomena. Hence, scientists concentrate on getting accurate data. As new observations are made and as technologies make possible more accurate measurements, new evidence can cause scientists to question their current explanations. This activity provides an opportunity for you to examine how an explanation changed over time.

Procedures and Questions

1. The following paper describes an experiment carried out by a scientist named J. B. Van Helmont that was published nearly four centuries ago (1662).
   a. Read the paper on this page and the next, underline Dr. Van Helmont's claim, and circle his summary of his evidence.

When a Willow Tree Grows, the Increase in Mass Comes Mainly From Water

Introduction

People have long wondered where the mass of a large tree comes from. Some have argued that the mass comes mainly from the soil. The experiment described below was designed to find an answer to the question of where the mass comes from when a giant tree grows from a tiny seedling.

Methods

I took 200 pounds of soil dried in an oven, moistened it with rainwater, and planted a 5 pound willow tree in it. Whenever the tree needed watering, I watered it with either rainwater or distilled water. After five years, I took out the tree, shook all of the soil out of its roots, and weighed it. I also dried all the soil the tree had grown in and weighed the soil again.

Results

After five years, the willow tree weighed 169 pounds and about three ounces and the dried soil weighed only a couple of ounces less than its former weight. The results are shown in Table 1.

Table 1. Results of Dr. Van Helmont's willow tree experiment

		Soil		Tree
Day 1	 =	200.0 lbs	+	 5.0 lbs
Five years later	 =	199.9 lbs	+	 169.2 lbs
Difference		-0.1 lbs		+164.2 lbs

Discussion and Conclusion

My experiment was designed to determine the source of the mass of a growing tree. I compared the increase in mass of the tree with the decrease in mass of the soil and found that the increase in the mass of the tree was far more than the decrease in mass of the soil. Since the only thing I added to the tree was water, the increase in mass must have come from the water.

b. Does Van Helmont's evidence support his conclusion? Why or why not? Does his explanation meet the elements of the Explanation Quality Criteria?

c. Write a better explanation for where most of the mass of a willow tree comes from. You may work with your partner or group to fill in the table below. However, once the table is complete, you should work independently to write your own explanation in the space provided on the next page.

Question	Where does most of the mass of a willow tree come from?
Answer	
Science Ideas	
Evidence	
Models	

Explanation:

Pulling It Together

1. Now that you have explained where the added mass of Van Helmont's willow tree came from, how would you answer the Key Question, **Where does the mass of growing plants come from?**

2. Can you think of other examples of a scientific explanation that has changed over time? What new evidence was found to support the revised explanation?

3. Are there other situations outside of science where using evidence-based explanations is important? If so, what are some examples?

Toward High School Biology

Chapter 4

Student Edition

Lesson 4.1—The "Stuff" That Makes Up Animals

What do we know and what are we trying to find out?

In the previous chapter you observed and modeled chemical reactions that contribute to plant growth. You figured out that plants grow by building carbohydrate polymers from glucose monomers they make from carbon dioxide and water molecules in their environment. You started by observing what body parts are getting bigger and heavier as plants grow and then looked at the molecules those body parts are made up of and the chemical reactions that produce those molecules. We will take a similar approach to understanding animal growth.

Answer the Key Question to the best of your knowledge. Be prepared to share your ideas with the class.

Key Question: What substances do animals need in order to grow and repair?

Activity 1: Observing Animal Growth

Different animals have different body structures. Some animals have wings, and others have arms. Some animals have feathers, and others have fur. Some animals have an internal skeleton, and others have an external skeleton. In this activity, you will observe what happens to body structures as animals grow.

Procedures and Questions

1. Observe the videos and photos of animal growth and repair. As you answer the questions below, think about the body structures animals have that you can see and also the body structures you can't see.

   a. What body structures are getting bigger as the **puppy** grows?

   b. What body structures are getting bigger as the **human girl** grows?

   c. What body structures are getting bigger as the **lobster** grows?

   d. What body structures are getting bigger or being rebuilt as the **lizard** regrows its lost tail?

   e. Do you think the repair of the lizard's lost tail is similar to growth? Why or why not?

   f. Where do you think the "stuff" comes from to make all of these body structures?

2. You have just observed that various animal body structures get bigger as they grow, but how do we know that animals are actually adding new matter to their bodies? To answer this question, we will observe what happens to the mass as animals grow and repair. Examine the data in Table 4.1 and answer the questions that follow.

Table 4.1. Typical Change in Mass (by Weight in Pounds or Grams) of Some Growing Animals, Over Different Time Periods

Animal	Initial Weight (Age)	Final Weight (Age)	Change in Weight	Increase or Decrease?
German shepherd dog (male)	20 pounds (at 2 months)	51 pounds (at 9 months)		
Human (female)	7.7 pounds (at birth)	64 pounds (at 10 years)		
Spiny lobster	3 pounds (at 11 months)	4 pounds (at 2 years)		
Lizard	0.62 grams (at time of tail loss)	1.72 grams (when new tail was fully grown)		

a. Using the data in Table 4.1, calculate the change in weight for each animal (you can use a calculator). Put a plus sign (+) next to any change in mass that was an increase. Put a minus sign (–) next to any change in mass that was a decrease.

b. Based on these data, what happens over time to the mass of an animal when it grows or repairs its body?

Activity 2: The Substances That Make Up Animals

As you have seen, animal bodies are very complex. They have a wide variety of body structures—but what makes up those structures? For example, you may have heard that 50%–60% of a human's weight is water. So what keeps us from being puddles on the ground? Animals, including humans, are made up of many other substances besides water, and these substances make up structures that give animals their shape. Aside from water, most of an animal's body is made up of three major types of substances: *proteins*, *fats*, and *carbohydrates*. You learned in Chapter 3 that although plants are made up of the same three major types of substances, plant body structures are made mostly of carbohydrates. Is this also true for animals? In this activity, you will find out how much of each type of substance makes up the bodies of various animals.

Procedures and Questions

1. Observe the photos and videos of the interior body structures of animals. Think about what they might be made of.

2. With your group, examine Tables 4.2 and 4.3 (p. 146) and answer the questions that follow.

    a. On Table 4.2, **circle** the type of molecule that is present in the highest amount in all the animals.

Table 4.2. Relative Mass (in Grams per 100 Grams) of Protein Molecules, Fat Molecules, and Carbohydrate Molecules in Some Animal Bodies

Animal	Protein (per 100 g)	Fat (per 100 g)	Carbohydrate (per 100 g)
Anchovy fish	65.3	7.1	4.4
Brine shrimp	58.4	4.0	29.3
Cow	60.0	8.9	15.1
Chicken	42.3	37.8	10.5
Crayfish	47.5	3.4	29.3
Hamster	49.8	34.6	8.1
Herring fish	72.7	8.5	0.8
Human	48.0	38.7	1.5
Locust	22.1	3.0	4.3
Squid	74.8	8.8	4.9
Worm	56.4	7.8	19.6

Table 4.3. Average Weight of Body Materials in a 154-Pound Human

Material	% of Body Weight
Muscle	41.5
Fat	18.0
Skeleton	15.8
Blood	8.0
Skin	6.9
All the rest (brain, liver, kidneys, etc.)	9.8

b. In the coming lessons, we will focus on muscle to understand animal growth. Based on the data in Table 4.3, why do you think it might be a good idea to focus on muscle?

c. Based on the data in Tables 4.2 and 4.3, what type of substance do you think animals will need to make the most of in order to build body structures for growth? Why do you think so?

Pulling It Together

Work on your own to answer these questions. Be prepared for a class discussion.

1. Now that you have seen how much mass animals gain when they grow, and the substances that make them up, how would you answer the Key Question, **What substances do animals need in order to grow and repair?** Provide evidence for your answer.

2. If mass is increasing as animals get bigger, what does that mean about the number of atoms making up their bodies?

3. Think back to the animal body structures that you have observed, such as hair, fur, skin, muscle, tendons, scales, and scar tissue. We just concluded that these body structures are mostly protein. Do you think all proteins are the same substance or different substances? Why or why not?

Lesson 4.2—Proteins in Animal Bodies and Food

What do we know and what are we trying to find out?

In the last lesson, you saw that animal body structures are largely made up of protein. Nutrition labels show the amount of protein in different foods, so we know that we can obtain protein from eating foods with protein, such as meat. Body builders say they need to eat a lot of protein in order to build muscles. Are all proteins the same substance? If different proteins are different substances, what makes them different?

You will learn about the proteins that make up animal bodies, as well as the proteins that animals eat when they eat meat. Of course, meat is actually some of the body structures of other animals.

Answer the Key Question to the best of your knowledge. Be prepared to share your ideas with the class.

> **Key Question: Are the proteins that animals eat exactly the same as the proteins that make up their own bodies?**

Activity 1: Comparing Proteins in an Animal's Food and Proteins That Make Up the Animal's Own Body

The food that an animal eats often looks quite different from the animal itself. In some cases, the differences are easy to observe. For example, a nursing puppy drinks only milk but needs to make new body structures as it grows. Some species of African snakes eat only eggs, yet they need to make body structures that look quite different from the egg. How is this possible? You know that animal body structures are mainly proteins, so it makes sense for us to examine the proteins that make up animal body structures and compare them with the proteins that make up an animal's food.

Procedures and Questions

1. Observe the photo of the nursing puppy and compare what the puppy eats with the body structures it needs to build.

2. Observe the video of the African snake eating an egg and the photo of the snake shedding its skin. Compare what the snake eats with the body structures it needs to build.

3. Examine Table 4.4 and use the information to answer the questions below.

Table 4.4. Several Proteins, Where They Are Found in the Body, and Some of Their Properties

Protein Name	Body Structures	Solubility in Water	Does It Form Fibers?
Casein	Milk	High	No
Collagen	Tendons, skin	Low	Yes
Hemoglobin	Blood	High	No
Keratin	Skin, scales, hair	Very low	Yes
Myosin	Muscle	Low	No
Actin	Muscle	Low	Yes
Ovalbumin	Egg white	High	No

a. Are milk and muscle made up of the same proteins? Use evidence to support your answer.

b. Are egg white and skin/scales made up of the same proteins? Use evidence to support your answer.

c. Can you think of any animals that consume mainly hemoglobin but need to make actin and myosin? List them.

d. Can you think of any animals that don't consume collagen but need to make it? List them.

e. Think about the fact that the human body is 50–60% water. Would proteins with high or low solubility in water be better for building human body structures? Explain.

f. Are all the proteins listed in Table 4.4 the same substance? Why or why not?

Activity 2: Understanding Proteins

In Activity 1, we saw that different proteins have different properties. We know that different substances have different properties because they are made up of different molecules. But if casein, hemoglobin, myosin, and collagen are all different substances, why do we call them all proteins? In this activity, you will examine the molecular structure of a piece of protein and use what you observe to explain why different proteins have different properties. This activity will also help you think about whether the proteins that are in an animal's food are the same as or different from the proteins that make up its body.

Procedures and Questions

In earlier chapters, you observed several different polymers and the monomers that make them up. You saw that both nylon and Kevlar polymers are made up of two different monomers. In the plant growth chapter, you saw that both starch and cellulose are made up of a single type of monomer—glucose. However, even though both starch and cellulose are made up of the same monomer, they have different properties because the glucose monomers are linked differently. In other words, simply changing the arrangement of the same monomers resulted in starch and cellulose having very different properties.

What about proteins? It turns out that proteins are also polymers. But there are thousands of different proteins. How is this possible? To figure it out, we need to examine the monomers—called amino acids—that make up protein polymers and how they are arranged in different proteins.

1. In Figures 4.1 and 4.2 (p. 152), color the O atoms red and the N atoms blue. Examine the two figures of molecules and answer the questions that follow.

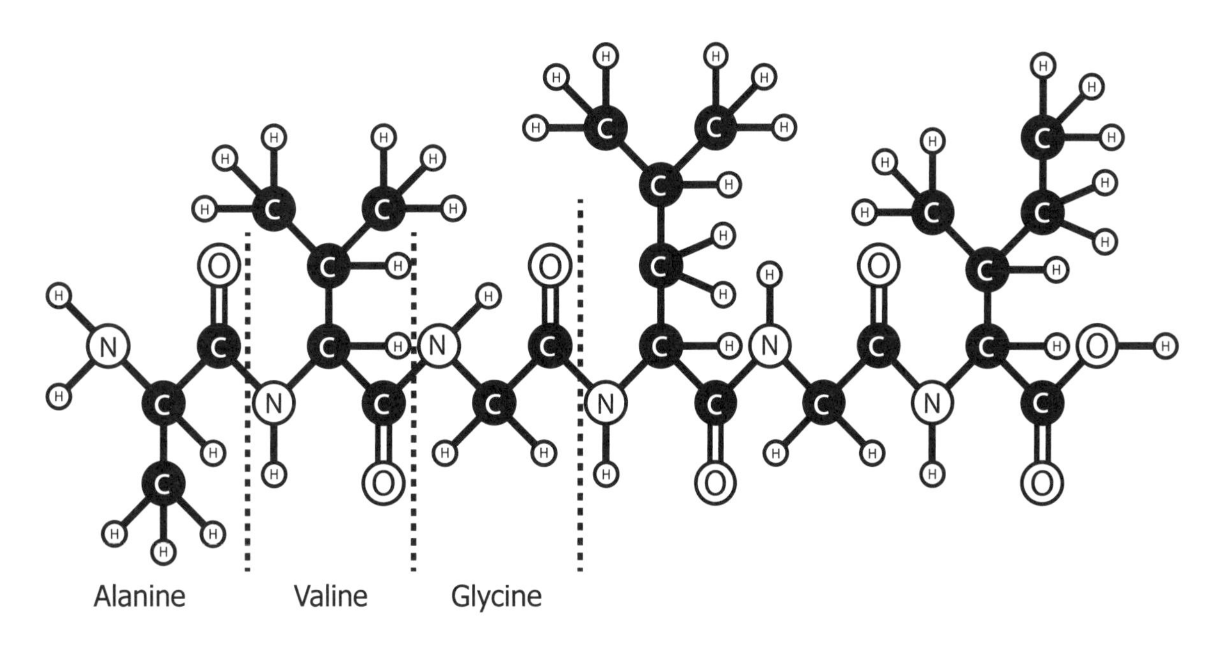

Figure 4.1. Model of the Structure of a Small Piece of a Protein Molecule

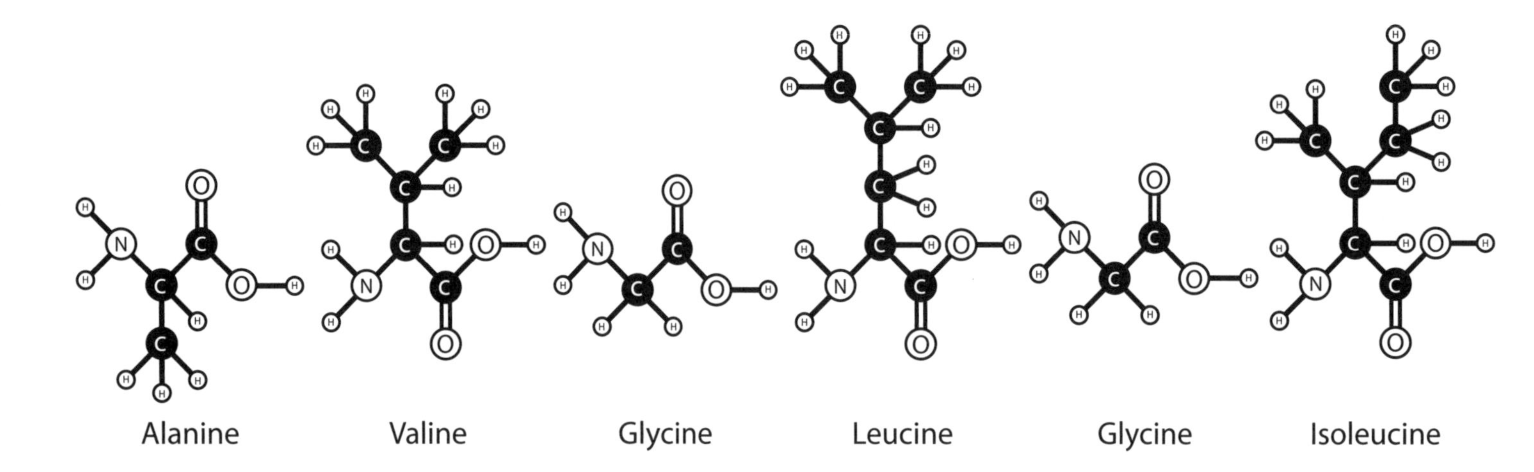

Figure 4.2. Flat Illustration of the Structure of the Amino Acids That the Piece of Protein in Figure 4.1 Would Be Made From

a. What do all the amino acids in Figure 4.2 have in common? Hint: Look at the left and right ends of each molecule.

b. What makes some of the amino acids in Figure 4.2 different from each other?

c. Now look at the piece of protein shown in Figure 4.1. The first three amino acid monomers have dotted lines between them and are labeled with their names (*alanine*, *valine*, and *glycine*). Can you find the others? Draw dotted lines between the last three amino acid monomers in Figure 4.1. Hint: Look for parts of the structure that are similar on the left and right ends of each monomer.

d. Do any of the amino acids in Figure 4.1 appear more than once in this piece of protein? If so, which one(s)? Hint: Do you see any from the first three somewhere in the last three?

e. Write the amino acid sequence of the piece of protein, using either the names of the amino acids or the first three letters of the names.

f. Now, write at least two other sequences that could be made up of the same amino acids.

g. We just looked at several amino acids. Most proteins are made up of 20 types of amino acid monomers that can be arranged to form thousands of different proteins. The protein ovalbumin in egg white is made up of 385 amino acid monomers, and actin, the protein found in muscle tissue, is made up of 374 amino acid monomers. Both proteins contain one or more of all 20 types of amino acid monomers. The sequence of the first 10 amino acids in each protein is shown below. (Only the first three letters of each amino acid name are shown.)

Ovalbumin (from hen egg white):
Met-Gly-Ser-Ile-Gly-Ala-Ala-Ser-Met-Glu ...

Actin (from rabbit muscle):
Asp-Glu-Thr-Glu-Asp-Thr-Ala-Leu-Val-Cys ...

Describe similarities and differences in the partial amino acid sequences of ovalbumin and actin.

h. We know that ovalbumin and actin have different properties. Explain, using atoms and molecules, why these two proteins have different properties.

Science Ideas

Lesson 4.2 was intended to help you understand some important ideas about proteins. Read the idea below. Remember that because science ideas are consistent with evidence from a wide range of examples, you are justified in applying them to new observations and data about similar examples. Look back through Lessons 4.1 and 4.2. Give at least one piece of evidence from your work so far that illustrates each part of this idea.

Science Idea #15: Proteins are the main polymers making up animal body structures. Protein polymers are molecules made of amino acid monomers. Different proteins have different properties because they are made of different types, numbers, and sequences of amino acid monomers.

Evidence:

Pulling It Together

Answer individually and be prepared for a class discussion.

1. Now that you've investigated a variety of proteins, how would you answer the Key Question, **Are the proteins that animals eat exactly the same as the proteins that make up their own bodies?** Answer the question and explain your answer using science ideas, evidence, and models.

2. Look back at the questions at the end of Activity 1. How do you think the egg-eating snake from Activity 1 might be able to get actin for building its muscles and keratin for building its scales from ovalbumin?

3. Most meats that humans eat are the muscles of other animals (cows, chickens, pigs, fish, and turkeys). When people prepare meat for meals, they typically use only the muscle. These muscles are made up mostly of the proteins *actin* and *myosin*. But when people grow, their bodies need to build more than just muscle. They need to build tendons (mostly collagen), skin (mostly collagen and keratin), and hair and nails (mostly keratin). What could explain how these different proteins could be made from actin and myosin?

Lesson 4.3—Explaining Animal Growth With Atoms and Molecules

What do we know and what are we trying to find out?

In the last lesson, we saw that the proteins an animal eats are rarely exactly the same proteins it needs to build its body structures. So what happens to food when an animal eats? When a puppy nurses, how does it get the proteins to build its body structures if it mainly takes in casein from its mother's milk? When the egg-eating snake ingests the protein ovalbumin, how does it get the proteins actin and myosin for its muscles and keratin for its skin?

In this lesson, you will have the chance to use experimental data and models to figure out what happens to the proteins in food once an animal has eaten them, and how they help an animal grow and repair its body structures.

Answer the Key Question to the best of your knowledge. Be prepared to share your ideas with the class.

> **Key Question: How do animals use proteins from food to build and repair their body structures?**

Activity 1: Observing What Happens to Food Proteins as an Animal Grows

In Chapter 3, you saw how scientists used labeled atoms to determine the reactants and products of chemical reactions that take place in plant body structures. The labeling technique was useful because scientists could prepare reactants with some labeled atoms and because the labeled atoms rearranged just like normal atoms do. You used data from these experiments to show that the carbon and oxygen atoms of the glucose molecules plants make come from carbon dioxide in the air. Do you think this technique can help us figure out how the egg-eating snake gets keratin, actin, myosin, and lots of other proteins when it eats mostly ovalbumin? Can it help us figure out how the puppy gets actin, myosin, hemoglobin, and lots of other proteins when it eats mostly casein in its mother's milk? In this activity, you will examine an experiment to help you answer these questions.

Procedures and Questions

1. Read about the experiment that scientists conducted. Examine the data they collected, and then answer the questions that follow with your team. Think about what the experiment can tell you about what happens to food when an animal eats. Observe the photos that your teacher will show you to help you understand the experiment.

Brine Shrimp Are a Suitable Food Source for Farming Herring

Introduction

Aquaculture is the farming of fish and other organisms that live in water, just as agriculture refers to the farming of land organisms. Scientists study fish in aquaculture to determine the best ways to grow the most fish. Part of this involves studying the food that is fed to the fish and how much fish grow when they are fed different foods. Scientists had previously found that catfish could be farmed by feeding them brine shrimp, tiny crustaceans that are relatives of shrimp people eat. In our study, we wanted to know if herring could also be farmed by feeding them brine shrimp.

Methods

We determined how to incorporate labeled carbon atoms into proteins in brine shrimp. We then fed these brine shrimp to young herring. We first measured how long the labeled carbon atoms from the brine shrimp proteins remained in the digestive system of each fish. Once we knew when those carbon atoms were out of the digestive system, we could see where they went. Data are shown in Figure 1.

Results

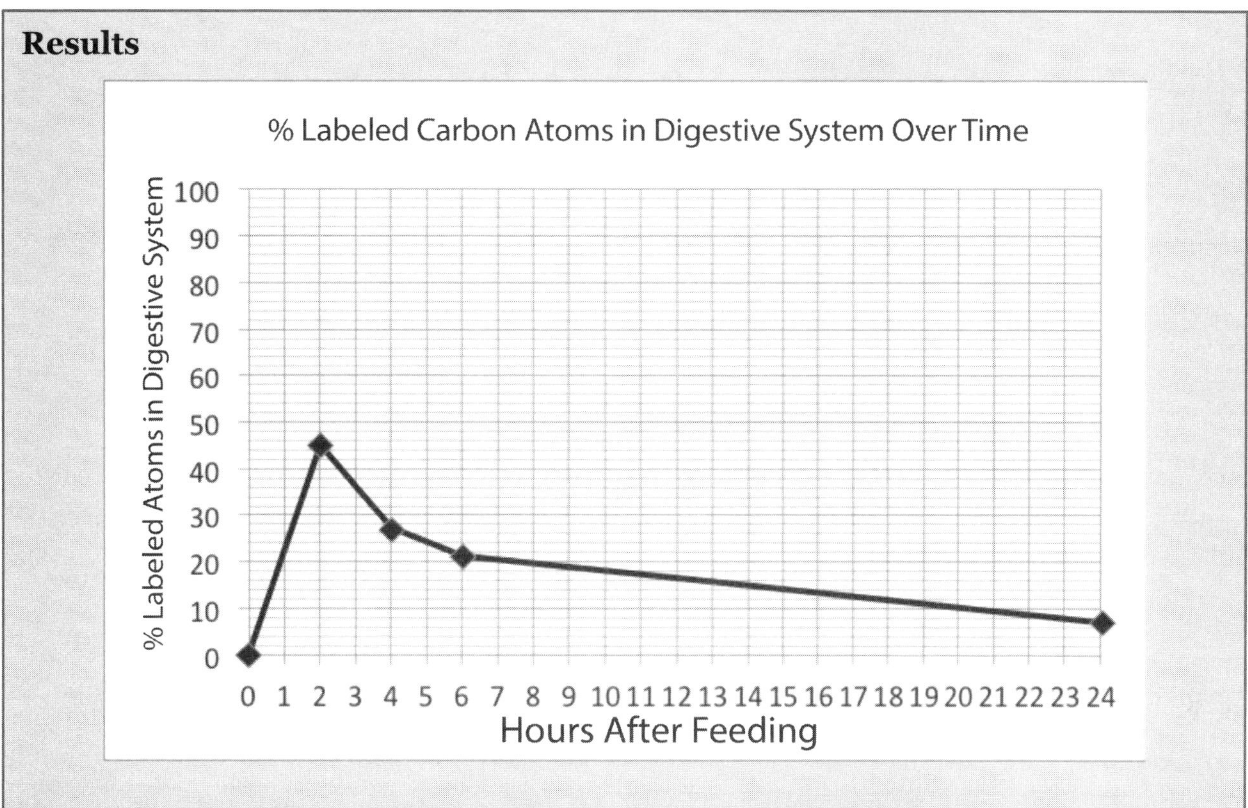

Figure 1. Percentage of labeled carbon atoms from food in a fish's digestive system over time (Time 0 was before feeding)

a. At which time points did the scientists take measurements? Circle them on Figure 1.

b. After feeding,

   i. when was the percentage of labeled atoms in the fish's digestive system the highest?

   ii. when was it the lowest?

c. What does Figure 1 tell you about the amount of food in the digestive system *over time*?

Twenty-four (24) hours after feeding, we measured the amount of labeled carbon atoms in three places: (1) in the water, (2) in the digestive system of each fish, and (3) in the rest of its body (as shown in Figure 2).

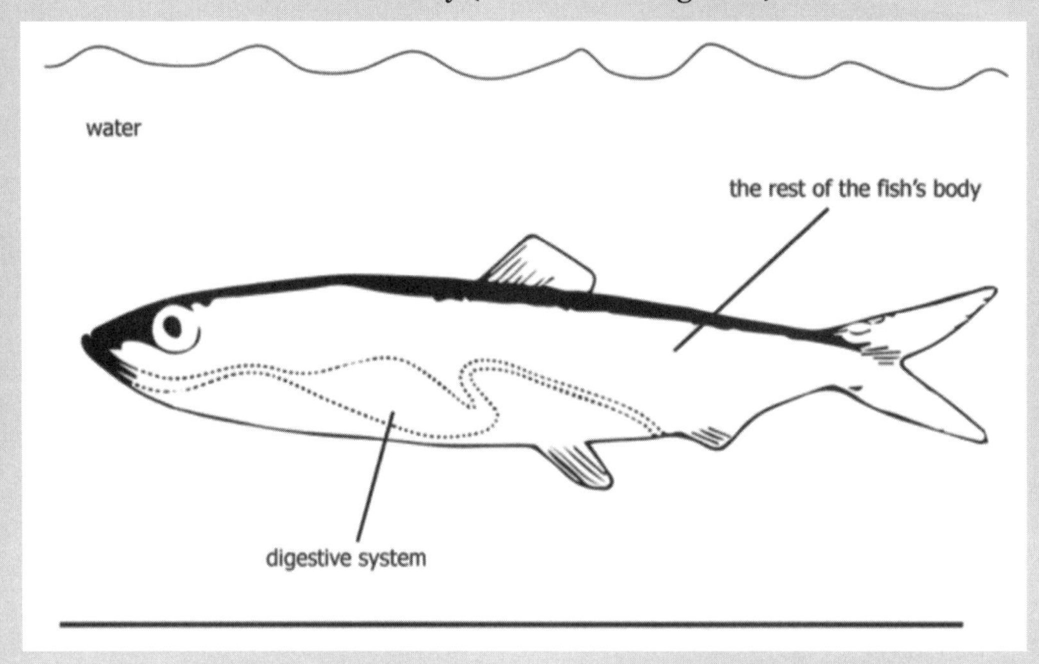

Figure 2. Representation of a herring in an aquaculture tank

Table 1. Percentage of labeled carbon atoms found in three locations 24 hours after feeding the fish with food containing labeled carbon atoms

Location	% of Labeled Carbon Atoms Found 24 Hours After Feeding
Digestive system	7
Rest of body (excluding digestive system)	20
Waste products (in the water, outside the fish's body)	73

Morais, S., L. E. C. Conceicao, M. T. Dinis, and I. Ronnestad. 2003. A method for radiolabeling Artemia with applications in studies of food intake, digestibility, protein and amino acid metabolism in larval fish. *Aquaculture* 231 (1–4): 469–487.

d. Remember that we are trying to find out what happens to the proteins in the food that an animal eats. What does Table 1 tell you about where the labeled atoms ended up once they left the fish's digestive system?

e. Did all of the labeled atoms end up as waste products? Give evidence for your answer.

f. When a fish eats brine shrimp, is it more like what happens when iron rusts in an open container or when baking soda reacts with vinegar in an open container? Explain your answer using science ideas from Chapter 2.

g. What do you think happened to the mass of the fish 24 hours after feeding—that is, 24 hours after time point 0? Why do you think so?

h. What will happen to the fish's mass if it continues to eat brine shrimp for many days?

Activity 2: Modeling the Breakdown and Building of Proteins in Herring

Materials

For each team of students

Prebuilt ball-and-stick models of four water molecules and a piece of a protein with a labeled carbon atom

Chemical Reaction Mats: Reaction #1: Breaking Down a Piece of a Protein into Amino Acids, Reaction #2: Building a Piece of a Protein from Amino Acids

Markers for writing on clear mat covers

Tray for working with models on desks

In Activity 1, you saw some data on where the carbon atoms in food go when it is eaten by an animal—in this case, herring. The data showed that some of the labeled carbon atoms ended up in the fish's body structures. You know that herring have a lot of muscle that they need to build in order to grow. Can the herring just take the protein from the brine shrimp and add it to their muscles unchanged? Or do chemical reactions play a role?

In Activity 2, we will use models to help answer that question. We will start with a piece of protein with a labeled carbon atom on a single amino acid and note where that amino acid is in the protein. Then we will model two chemical reactions—protein digestion and protein synthesis—that are involved when animals take in and use food to build and repair their bodies. Our models will help us see whether the labeled amino acid in the reactant molecules ends up in the same place in the amino acid sequence of the product molecules.

Procedures and Questions

Complete the steps in the procedure and respond to the questions. Steps 1–8 focus on the process by which proteins in food are digested in an animal's body. Steps 9–16 focus on the process by which new proteins are made (synthesized).

1. As a team, examine your model of a piece of food protein. Find the carbon atom that has a sticker on it. The sticker represents the label on the carbon atom. Write down the amino acid sequence of the piece of protein and highlight the amino acid that has the label. Where in the amino acid sequence is the labeled amino acid?

2. Look at the yellow side of the *Chemical Reaction Mat* for Reaction #1, which shows what the products of the chemical reaction will be. How many amino acids will this piece of food protein break down into?

3. How many connections between atoms in the food protein will need to break in order to make the amino acids?

4. Using the markers provided, draw lines on the green side of the mat through the connections your team plans to break in the piece of food protein. (Ask your teacher for help if you are having trouble deciding. All teams should check with their teacher before moving on to Step 5.)

5. Assign each team member a connection to break. Put your initials by the line through the piece of food protein that you will be breaking.

6. Flip your *Chemical Reaction Mat* to the yellow side (*Products*), remove the yellow sheet from the plastic sleeve, and place it next to the green side (*Reactants*).

7. Using only your piece of food protein and water molecules, build models of the products as shown on the yellow side of the mat. After you've finished building the products, **do not take apart your models.** Put the yellow sheet back in the sleeve.

8. With your team, answer the following questions.

   a. You started with a small piece of a protein polymer and four water molecules. What happened to the water molecules during the chemical reaction?

   b. Now that you've made the products, are any atoms connected to different atoms than they were before? If so, give an example.

c. Is the label on the same carbon atom as at the start of the chemical reaction? Is the labeled carbon atom part of the same molecule as it was at the start of the chemical reaction?

d. Do you think a protein and an amino acid will have the same properties? Why or why not?

e. Did you gain any atoms or lose any atoms during this chemical reaction?

f. Write a word equation for the reaction you just modeled.

g. Where in the fish's body do you think Reaction #1 happens?

9. Your teacher will give you a new *Chemical Reaction Mat* for Reaction #2.

10. With your team, examine Reaction #2. The products you made in Reaction #1 should match the pictures on the green side of the mat for Reaction #2.

11. Remove the yellow side of the mat for Reaction #2, which shows the products of the chemical reaction, and place it next to the green side. Using the green side to identify the amino acids, write down the amino acid sequence of the protein product (yellow side).

12. Arrange your amino acid models according to the sequence you wrote down, and have your teacher check the arrangement.

13. On the green side of the mat for Reaction #2, circle the atoms that will separate and form new connections.

14. Assign each team member a connection to make. Put your initials by atoms that you will separate and use to form new connections.

15. Build the products for Reaction #2 as shown on the yellow side of the mat. Place the yellow side of the mat back in the sleeve. **Do not take apart your models of the polymer—leave them for the next class.**

16. With your team, answer the following questions.

    a. Write down the amino acid sequence of the piece of protein you just built, and highlight the amino acid that has the label. Where in the sequence is the labeled amino acid?

    b. Compare the amino acid sequence of the model of the protein you just produced with the amino acid sequence of the model of the protein you started with in Step 1. Are the proteins you modeled the same or different? Use science ideas and models to explain your answer.

    c. Where in the fish's bodies would the products from Reaction #2 be found? Support your answer with evidence.

d. Let's consider how the two chemical reactions you modeled relate to the herring data. Where were the atoms at the beginning of the experiment (time 0 on the graph)? Where were the atoms after 24 hours?

e. What does your answer for Step 16d mean for the measured mass? Will it change or stay the same? Why?

Science Ideas

Lesson 4.3 was intended to help you understand some important ideas about how food relates to growth in animals. Read the ideas below. Notice that Science Ideas #16 and #17 explain observations about animal growth in terms of atoms. We can observe when animals grow, but we can't see atoms rearranging or entering or leaving the system. However, because Science Ideas #16 and #17 state general principles that are consistent with a wide range of observations and data, we can use them to reason about the growth of all animals. You will be expected to use ideas about atoms to explain phenomena involving animal growth.

Look back through Lesson 4.3. Give at least one piece of evidence from your work so far that illustrates each of the science ideas listed below.

Science Idea #16: The process by which proteins from food become part of animals' body structures involves chemical reactions in which the proteins from food are broken down into amino acid monomers, and these monomers are used to build different protein polymers that make up body structures. Atoms are rearranged during both the breakdown and the building of protein polymers.

Evidence:

Science Idea #17: When animals grow or repair, they increase in mass. Atoms are conserved when animals grow: The increase in measured mass comes from the incorporation of atoms from molecules that were originally outside of the animals' bodies.

Evidence:

Pulling It Together

1. After examining the data and models in this lesson, how would you answer the Key Question, **How do animals use proteins from food to build and repair their body structures?**

2. We used models to reason about how a fish makes its body protein from eating brine shrimp. Now explain, using models, how egg-eating snakes can make proteins like myosin and keratin when they mainly eat ovalbumin.

3. Imagine your friend has just finished eating a turkey sandwich. You can represent the atoms making up proteins in the turkey meat with LEGO bricks. How would the total number of bricks that were in the sandwich before your friend ate it compare with the total number of bricks inside your friend's body immediately after she ate it? Would all the bricks stay inside her body? Would any of the bricks disappear? Support your answer with evidence.

4. When you get a cut on your hand, your body builds scar tissue to seal up the cut. Scar tissue is made up of the protein *collagen*. Explain how this can happen. Be sure your explanation meets the Explanation Quality Criteria.

5. Describe how the addition of new protein molecules contributes to the growth of an animal's body structures and body.

Lesson 4.4—Examining Explanations of Animal Growth and Repair

What do we know and what are we trying to find out?

In Chapter 4, you have investigated how animals grow. You have applied what you learned about atoms, molecules, chemical reactions, and mass in Chapters 1, 2, and 3 to explain animal growth. This lesson will give you a chance to examine and evaluate other people's explanations of animal growth.

Answer the Key Question to the best of your knowledge. Be prepared to share your ideas with the class.

Key Question: How are the explanations we have been writing similar to the explanations that scientists write when they publish their work?

Activity 1: Analyzing Scientists' Explanation About Animal Growth and Repair

Scientists tell others about new research findings by presenting them at meetings and publishing them in scientific journals. In order to get their work published in well-respected journals, scientists must state their conclusions and explain them with evidence and logical arguments. However, scientists often don't use the terms *evidence* or *science ideas* in their papers because other scientists know how to recognize the elements of an explanation. This activity will help you learn to recognize these elements in a published paper (and sometimes to recognize when elements are missing).

Procedures and Questions

1. Read the simplified paper below and answer the questions that follow with your group.

Taking an Amino Acid Supplement Causes Elderly Men to Increase Protein Production

Introduction

During aging, a gradual decrease in skeletal muscle mass occurs in both rodents (like mice and rats) and humans. This decrease in muscle mass often makes elderly people weaker and affects their ability to move around and their health. Prior studies have shown that when elderly rats were fed high amounts of an amino acid supplement, the rats had more muscle mass and produced more muscle protein (myosin) than similar rats that did not get the amino acid supplement. Scientists wanted to know if the amino acid supplement would also help elderly men produce more muscle protein and if that could stop or slow down the decrease in muscle mass.

Methods

A group of 70-year-old male volunteers were randomly assigned to either an **experimental** or a **control** group. The volunteer men were approximately the same size and weight and in good health. "The nature and potential risks of the study were fully explained to each volunteer and written informed consent was obtained before the study from each participant" (p. 306).

The methods were approved by the local ethical committee. The experimental group was fed complete meals (containing proteins, carbohydrates, and fats) plus an amino acid supplement. The control group was fed the complete meals without the amino acid supplement.

The amount of myosin protein produced was measured by giving all volunteers a tiny amount of labeled phenylalanine (an amino acid found in myosin) and measuring how much myosin-containing labeled phenylalanine each man's body produced. High amounts of labeled myosin would indicate high levels of protein synthesis, whereas low amounts of labeled myosin would indicate low levels of protein synthesis.

Results

Table 1. Myosin Production in the Presence and in the Absence of Amino Acid Supplement

	What They Ate	Amount of Labeled Myosin (Muscle Protein)
Experimental Group	Complete meals plus amino acid supplement	High
Control Group	Complete meals	Very low

Discussion and Conclusion

Our study was designed to assess the impact of amino-acid-supplemented meals on protein synthesis in elderly volunteers. We compared the amount of labeled myosin produced by volunteers fed complete meals plus an amino acid supplement with that produced by volunteers fed only the complete meals. We showed that supplementing a complete diet with the amino acid increased muscle protein synthesis. Additional experiments would be necessary to find out whether increasing protein synthesis will actually reduce or prevent the decline in muscle mass of elderly people.

Source: Rieu, I., M. Balange, C. Sornet, C. Giraudet, E. Pujos, J. Grizard, L. Mosoni, and D. Dardevet. 2006. Leucine supplementation improves muscle protein synthesis in elderly men independently of hyperaminoacidaemia. *Journal of Physiology* 575 (1): 305–315.

a. What is the authors' conclusion?

b. What evidence do the authors present?

c. Why do you think the authors included the fact that the volunteers were approximately the same size and weight and in good health in the paper?

d. The simplified paper left out a lot of information about the methods that the scientists used in their study. The original paper included details about what participants were fed, where the supplements were purchased, and how the amount of labeled myosin was determined. What do you think is the purpose of including all that information in the original paper?

e. The simplified paper also left out references to earlier studies. For example, the original paper cited previous findings showing that when elderly rats were fed high amounts of an amino acid supplement, the rats had more muscle mass and produced more muscle protein (myosin) than similar rats that did not get the same supplement. What do you think is the purpose of mentioning the studies with rats?

Pulling It Together

1. After examining the scientists' explanation, how would you answer the Key Question, **How are the explanations we have been writing similar to the explanations that scientists write when they publish their work?** Be sure to talk about the criteria that both you and scientists would use for evaluating the quality of an explanation.

2. A friend says in class that chemistry has nothing to do with biology. He says that "chemical reactions only happen in labs, not in people or animals!" After some discussion, the class comes up with the statements below. In the space next to each statement, provide evidence, science ideas, or models to support it.

Statements that will help explain your answer	Evidence, science idea, or modeling activity that supports each statement
Animals increase in mass as they grow.	
Animal body structures are made up mostly of protein.	
Animals make body structures from their food.	
The proteins that make up animal body structures are different from the proteins in their food.	
Chemical reactions are required to make new protein molecules for building body structures.	
It is possible to make a body protein by digesting a food protein to amino acids and reordering the amino acids to make a different protein.	

Lesson 4.5—Explaining Growth in All Living Things

What do we know and what are we trying to find out?

In Chapters 3 and 4, you investigated how plants and animals grow and applied ideas about atoms and molecules to explain how they produce new substances and increase in mass. This final lesson gives you a chance to identify things that plant growth and animal growth have in common and to apply your understanding to a different type of living thing: mushrooms.

Key Question: How does what we have learned about the growth of plants and animals help us think about the growth of all living things?

Activity 1: Venn Diagram: Animal Growth and Plant Growth

Some things about animal and plant growth are similar, while some things are different. In this activity, you will create a visual, called a Venn diagram, which will help you summarize your ideas about these similarities and differences.

Procedures and Questions

1. Review the before and after photos of animal growth and plant growth that your teacher shows you.

2. Use the template provided by your teacher to create a Venn diagram. Work on your own.

   a. Include as many similarities and differences as you can. You should have at least five items for both, and at least three items each for "only plant growth" and "only animal growth."

   b. Think about both things you can see and things you can't see with the human eye.

   c. Include what happens to mass and what happens to atoms in each of the processes.

   d. Feel free to look through your whole notebook for ideas.

Tape Venn diagram here.

Activity 2: Growth in Mushrooms

In this activity, you will apply everything that you have learned about chemical reactions and growth in animals and plants to a different kind of living organism: mushrooms.

Procedures and Questions

1. Read about mushrooms, observe the photos your teacher will show you, and work with your group to answer the question in Step 1a. Then, answer the question in Step 1b individually.

- Some living organisms are neither plants nor animals. You have probably heard of other groups of living things, such as bacteria, algae, and fungi. Fungi are a diverse group of living things, including a wide variety of mushrooms.
- Mushrooms do not carry out photosynthesis like plants do. They typically get the monomers they need to build their body structures from dead or decaying organisms or other biological materials. They can digest the dead organisms and absorb the monomers. That is why mushrooms are often found attached to the trunk of a dead or dying tree. Trees are made up mostly of cellulose.
- Most mushrooms have the same basic body structures, including a cap that produces spores for reproduction, a stem that holds up the cap, and a threadlike feeding network that takes in nutrients. Mushroom body structures are made up largely of chitin (pronounced "ki-tin") polymers. Chitin is another type of carbohydrate polymer like the cellulose and starch you saw earlier.

   a. Make a list of chemical reactions that could help explain how a mushroom grows and gains mass. Next to each statement, list evidence and science ideas from Chapters 3 and 4 that support each statement. Use the table on the next page to organize your thinking.

Statements about chemical reactions that could help explain how a mushroom grows and gains mass	Evidence and science ideas that support each statement

b. If mushrooms are growing on a fallen dead tree, what do you predict will happen to the mass of the fallen tree as the mushrooms continue to grow? Will the mass of the fallen tree increase, decrease, or stay the same? Does your prediction violate ideas about conservation of matter? Use science ideas from anywhere in the unit to explain your prediction.

Pulling It Together

1. Now that you have completed the unit, how would you answer the Key Question, **How does what we have learned about the growth of plants and animals help us think about the growth of all living things?**

2. How are changes in the matter that makes up living and nonliving things similar?